TOWARD A COORDINATED SPATIAL DATA INFRASTRUCTURE FOR THE NATION

Mapping Science Committee
Board on Earth Sciences and Resources
Commission on Geosciences, Environment and Resources
National Research Council

National Academy Press
Washington, D.C. 1993

NOTICE: The project that is the subject of this report was approved by the Governing Board of the National Research Council, whose members are drawn from the councils of the National Academy of Sciences, the National Academy of Engineering, and the Institute of Medicine. The members of the committee responsible for the report were chosen for their special competences and with regard for appropriate balance.

This report has been reviewed by a group other than the authors according to procedures approved by a Report Review Committee consisting of members of the National Academy of Sciences, the National Academy of Engineering, and the Institute of Medicine.

The National Academy of Sciences is a private, nonprofit, self-perpetuating society of distinguished scholars engaged in scientific and engineering research, dedicated to the furtherance of science and technology and to their use for the general welfare. Upon the authority of the charter granted to it by the Congress in 1863, the Academy has a mandate that requires it to advise the federal government on scientific and technical matters. Dr. Frank Press is president of the National Academy of Sciences.

The National Academy of Engineering was established in 1964, under the charter of the National Academy of Sciences, as a parallel organization of outstanding engineers. It is autonomous in its administration and in the selection of its members, sharing with the National Academy of Sciences the responsibility for advising the federal government. The National Academy of Engineering also sponsors engineering programs aimed at meeting national needs, encourages education and research, and recognizes the superior achievements of engineers. Dr. Robert M. White is president of the National Academy of Engineering.

The Institute of Medicine was established in 1970 by the National Academy of Sciences to secure the services of eminent members of appropriate professions in the examination of policy matters pertaining to the health of the public. The Institute acts under the responsibility given to the National Academy of Sciences by its congressional charter to be an adviser to the federal government and, upon its own initiative, to identify issues of medical care, research, and education. Dr. Kenneth I. Shine is president of the Institute of Medicine.

The National Research Council was organized by the National Academy of Sciences in 1916 to associate the broad community of science and technology with the Academy's purposes of furthering knowledge and of advising the federal government. Functioning in accordance with general policies determined by the Academy, the Council has become the principal operating agency of both the National Academy of Sciences and the National Academy of Engineering in providing services to the government, the public, and the scientific and engineering communities. The Council is administered jointly by both Academies and the Institute of Medicine. Dr. Frank Press and Dr. Robert M. White are chairman and vice chairman, respectively, of the National Research Council.

Support for this study by the Mapping Science Committee was provided by the Defense Mapping Agency, the United States Geological Survey, the Bureau of Land Management, and the Bureau of the Census.

Library of Congress Catalog Card No. 93-84335
International Standard Book No. 0-309-04899-0

Copies of this report are available from
National Academy Press
2101 Constitution Avenue
Washington, D.C. 20418

B-149

Printed in the United States of America

MAPPING SCIENCE COMMITTEE

JOHN D. BOSSLER, The Ohio State University, *Chairman*
JOHN C. ANTENUCCI,[a] PlanGraphics, Inc.
LAWRENCE F. AYERS, Intergraph Corporation
BARBARA P. BUTTENFIELD,[b] State University of New York, Buffalo
ROBERT LEE CHARTRAND, Naples, Florida
DONALD F. COOKE, Geographic Data Technology, Inc.
DAVID J. COWEN,[a] University of South Carolina
JOHN E. ESTES,[a] University of California, Santa Barbara
LEE C. GERHARD,[b] Kansas Geological Survey
MICHAEL F. GOODCHILD,[b] University of California, Santa Barbara
CLIFFORD GREVE,[c] Autometrics, Inc.
GIULIO MAFFINI, Intera-Tydac
JOHN D. MCLAUGHLIN, University of New Brunswick
BERNARD J. NIEMANN, JR., University of Wisconsin, Madison
BARBARA B. PETCHENIK, R.R. Donnelley & Sons Company (deceased, June 1992)
GERARD RUSHTON, University of Iowa
HOWARD J. SIMKOWITZ,[a] Caliper Corporation
LARRY J. SUGARBAKER,[b] Washington State Department of Natural Resources
ROBERT TUFTS, The Analytic Science Corporation

NRC Staff
THOMAS M. USSELMAN, Senior Staff Officer
JUDITH ESTEP, Administrative Secretary

[a] Term ended May 31, 1992
[b] Term began June 1, 1992
[c] Resigned December 1991 when he began job with the U.S. Geological Survey

BOARD ON EARTH SCIENCES AND RESOURCES

COMMISSION ON GEOSCIENCES, ENVIRONMENT, AND RESOURCES

PREFACE

The Mapping Science Committee (MSC) was established in 1987 at the request of the U.S. Geological Survey (USGS) to provide advice on their cartographic and geographic activities. During the course of its studies, the MSC was exposed to activities in several other federal agencies and recognized several generic problems involving the conduct of mapping and spatial analysis. Because of the continuing explosive growth of the technology and the accompanying modernization efforts (both within the federal and nonfederal government and elsewhere), generic problems, the redundancy of data production, the potential for application of the methodologies to other programs, and the large fiscal expenditures anticipated, the MSC expanded its scope to offer its capabilities and advice to other federal agencies that have become involved in these programs.

In response to the initial charges developed for the earlier USGS activity, the MSC issued two reports, *Spatial Data Needs: The Future of the National Mapping Program* (1990) and *Research and Development in the National Mapping Division, USGS: Trends and Prospects* (1991). In both reports, the MSC recognized the utility of its advice to the broader governmental agencies that utilize mapping or the analyses of spatially referenced digital data.

Because of this recognition, the MSC initiated this study. The question that this study addresses is: What could be done better or more efficiently if the content, accuracy, organization, and control of spatial data were different? The study examines the current national spatial data infrastructure, encompassing the roles of private institutions as well as local, state, and federal governments in using and sharing geographic information. In addition it identifies barriers that prevent these groups from acquiring knowledge, making decisions, or performing the duties that rely on the timely availability and easy access to an organized body of geographically referenced information. The scope of spatial data can be enormous, and spatial data can be important components of a wide variety of scientific, technical, and social disciplines and applications. The MSC focused its efforts on the generic issues of spatial data management, collection, and use, particularly regarding the data bases that drive geographic information systems and other similar methods of analyses.

In the past 2 years, the MSC met (at least once) with 18 different federal agencies to be briefed and to discuss various programs within each agency dealing with spatial data collection and use. These agencies include the following:

- Department of Commerce
    - Bureau of the Census
    - National Institute of Standards and Technology
    - National Oceanic and Atmospheric Administration
- Department of Defense
    - Army Corps of Engineers
    - Defense Mapping Agency
- Department of the Interior
    - Bureau of Land Management
    - Fish and Wildlife Service
    - U.S. Geological Survey
- Department of Agriculture
    - Agricultural Stabilization and Conservation Service
    - Economic Research Service
    - Forest Service
    - National Agriculture Statistical Service
    - Soil Conservation Service
- Department of Transportation
    - Federal Highway Administration
- Environmental Protection Agency
- National Aeronautics and Space Administration

At several MSC meetings, discussions of policies regarding spatial data were held with representatives of the Office of Management and Budget and its interagency committee, the Federal Geographic Data Committee (FGDC). In addition, representatives of the MSC attended several FGDC meetings. Representatives from Bell South also participated in a meeting and discussed issues relevant to the utilities sector, and a representative from the Council of State Governments discussed the Council's *State Geographic Information Activities Compendium*. The MSC appreciates the participation of officials from these organizations in developing the committee's background to address the issues in this report.

The MSC, through its members, brought valuable experience relevant to state and local governmental activities, the needs of utilities, Intelligent Vehicle Highway Navigation Systems, and the role of the private sector. Members of the MSC participated in key roles at the 1991 FGDC-sponsored Geographic Information and Spatial Data Exposition held in Washington, D.C., and at other meetings of associations and societies relevant to spatial data.

CONTENTS

1 EXECUTIVE SUMMARY 1
THE CHALLENGE 1
ISSUES 3
RECOMMENDATIONS 5
CONCLUSION 6

2 INTRODUCTION 7
REFERENCE 11

3 NATIONAL SPATIAL DATA INFRASTRUCTURE 12
EVOLUTION 12
Stage I (1960—1980), 14; Stage II (1975—2000), 14; Stage III (1990 and Beyond), 15
CONCEPT 15
REFERENCES 18

4 CURRENT SITUATION 19
MAJOR ISSUES 20
FEDERAL AGENCY ACTIVITIES 27
Office of Management and Budget, 29; Federal Geographic Data Committee, 30; Other Federal Agencies, 31
STATE AGENCIES 43
Current Situation, 43; Federal Relationships, 45; Standards, 50; Problems, 50
LOCAL GOVERNMENTS 50
Current Situation, 51; Federal Relationships, 52 Standards, 52; Problems, 52
PRIVATE SECTOR 53
ACADEMIA 56
REFERENCES 57

5 SPATIAL DATA AND THE URBAN FABRIC 59
INTRODUCTION 59
BACKGROUND 60
LAND BASE SYSTEMS COMPONENT 61
Description, 61; Primary Responsibility, 62; Needs, 63
Models, 63; Rationale for Federal Involvement, 63
THE NATION'S CADASTRE 64
Introduction, 64; Description, 65; Primary Responsibility, 66; Rationale for Federal Involvement, 66;
The Problems, 66; Responding to the Problems, 67
STREET CENTERLINE SPATIAL DATA BASE 68
Description, 68; Significance and Applications, 68;
Current Status of Street Centerline Spatial Data Bases, 69; Problems Caused by Deficiencies of Available SCSDs, 70; Need for Greater Coordination or Consolidation , 71; Conclusions: The Need to Ensure Access, Use, and Maintenance, 72
REFERENCES 73

6 SPATIAL DATA AND WETLANDS 74
INTRODUCTION 74
EVALUATION PROCESS 78
Public Interest in Wetland Protection, 78; Lack of a Collective Perspective, 79; Impediments, 80;
Wetland Information Diffusion Model, 85
CONCLUSION 86
REFERENCES 88

7 SHARING OF SPATIAL DATA 89
RATIONALE FOR A DATA SHARING PROGRAM 89
Objectives, 89; Examples of Data Sharing Programs, 90;
A PROPOSED SPATIAL DATA SHARING PROGRAM . . 94
Key Concepts, 94; NSDI Spatial Data Sharing Program, 98;
Guidelines for System Implementation, 99
MECHANISMS FOR IMPLEMENTING A SPATIAL DATA SHARING PROGRAM 104

SPATIAL DATA CATALOGS 106
Distributed Data Catalogs, 106; Applications , 109
REFERENCES 110

8 CONCLUSIONS 111
IMPROVING THE NSDI 112
The Principles, 112; The Components, 113
A NEW STRATEGY 116
Obtain and Maintain National Commitment, 116; Evaluate Requirements, Constraints, and Opportunities, 117; Determine Priorities, 117; Develop Coordination Mechanisms and Organizational Structures, 117; Assign Roles and Responsibilities, 118; Develop Standards and Policies 118; Develop and Monitor Projects, 118; Identify and Resolve Issues, 118
REFERENCES 119

9 RECOMMENDATIONS 120

APPENDIX A: SPATIAL DATA AND WETLANDS 127
INTRODUCTION 127
THE NATURE OF WETLANDS 129
STATE AND CONDITION OF WETLANDS 132
PROTECTION OF WETLANDS 134
INFORMATION REQUIREMENTS 137
IMPEDIMENTS TO A NATIONAL WETLAND INFORMATION SYSTEM 141
INFORMATION DIFFUSION AND EVOLUTION 155
CONCLUSION 163
REFERENCES 165

ACRONYMS 169

This volume is dedicated to the memory of Barbara Bartz Petchenik (1939—1992), a friend and colleague, whose inspiration, contagious enthusiasm, and keen insight provided profound intellectual stimuli to the Mapping Science Committee. Barbara had been associated with the committee since its inception in 1987 and made significant contributions to all three of the committee's reports. Her bright smile and refreshing perspective will be missed by all.

TOWARD A COORDINATED SPATIAL DATA INFRASTRUCTURE FOR THE NATION

1
EXECUTIVE SUMMARY

THE CHALLENGE

Rapid access to data and information is crucial to the economic, environmental, and social well-being of our global society. It is generally accepted that in the early 1960s the United States was moving towards being an information society. This information society depends on spatial (geographic) data and information. Today an ever increasing volume of these spatially referenced data are being produced, stored, transferred, manipulated, and analyzed in digital form. Until now, maps in analog form have been a mainstay of a wide variety of applications and decision making. This is changing as more data and information on a wider variety of topics or themes (e.g., population, hydrology, agriculture, climate, and soils) become available in digital format.

To service those who need digital data, new digital products are appearing with greater frequency, increasing quantities of spatially referenced data. With this increased production comes the potential for substantial duplication of effort or the underutilization of valuable information that may have been created at considerable cost and effort.

The scope of spatial data can be enormous, and spatial data can be important components of a wide variety of scientific, technical, and social disciplines and applications. In the context of this study the Mapping Science Committee focused its efforts on the generic issues of spatial data management, collection, and use, particularly in geographic information systems and other similar methods of analyses.

A major challenge over the next decade will be to enhance the accessibility, communication, and use of spatially referenced data to support a wide variety of decisions at all levels of society. By creating an effective, efficient, and widely accessible "information highway"—the backbone of a robust National Spatial Data Infrastructure (NSDI)—data could be readily transported and easily integrated both horizontally (e.g., across environmental, economic, and institutional data bases) and vertically (e.g., from local to national and eventually to global levels). The NSDI could provide transparent access to myriad data bases for countless applications (e.g., facility management, real estate transactions, taxation, land-use planning, transportation, emergency services, environmental assessment and monitoring, and research). Work on these applications occurs in schools, offices, and homes nationwide. Furthermore, a robust NSDI will create new value-added services and market opportunities in emerging spatial information industries.

The National Spatial Data Infrastructure is the means to assemble geographic information that describes the arrangement and attributes of features and phenomena on the Earth. The infrastructure includes the materials, technology, and people necessary to acquire, process, store, and distribute such information to meet a wide variety of needs.

We must emphasize that a national spatial data infrastructure *exists*. It is an *ad hoc* affair because, until very recently, no one conceived of it or defined it as a coherent entity, and indeed it has not been very coherent or coordinated. It is not the task of the Mapping Science Committee (MSC) to *create* a national spatial data infrastructure. We want merely to point out its existence, identify its components and characteristics, assess the efficiency with which it functions to meet national needs (from local to federal), and finally make recommendations that might make it more useful, more economical, better coordinated, and robust. Several investigators have shown that investments in spatial data technologies are normally repaid by the long-term benefits and that greater efficiencies are realized. In addition, there could be a significant reduction in the cost of operation of geographic information systems (GIS) if existing data were shared, thus reducing duplication of efforts of data collection. The committee maintains that improvements in the national spatial data infrastructure are critical to the maintenance of a competitive position for the United States in an increasingly international economic arena.

Over the past decade researchers in government, private industry, and academia began to appreciate the technical and institutional difficulties in creating distributed networks of spatial data bases at all levels of government and society. Many technical problems have been solved, but most networks in place today are still at a primitive stage of development and involve only a limited number of organizations and data types. Even when the number and the size of data bases involved in a given network are relatively small, the coordination and cooperation required for data sharing have often been hard to obtain. Policy issues such as levels of incentives, mandates, access, pricing, privacy, and liability have only recently begun to be addressed.

The nation's need to access spatial data and information is growing rapidly. Geographic referencing is needed in areas such as health, education, and social welfare, where a variety of information collected from many sources is used to track problems and identify trends. Perhaps, the most rapidly growing requirements for spatial data and information is currently in environmental management. To achieve such goals as sustainable economic development and protection of sensitive natural resources (e.g., wetlands), land managers need to know what information exists, how to obtain it, and how it can be merged with information from other sources. New technologies (e.g., GIS, remote sensing, spatial modeling, and artificial intelligence) provide the capability to meet these and other needs. However, unless the National Spatial Data Infrastructure is robust and the spatial data bases, policies, and standards are in place to facilitate the access and use of spatial data nationally, opportunities in areas from environment to development will be lost.

An enormous amount of resources is expended annually in spatial data systems, data collection, and manipulation. The annual amount is difficult to quantify. Several marketing firms have estimated some of the worldwide expenditures: roughly, there are annual expenditures in related software sales of about $600 million and hardware sales of about $1,300 million in 1992. From the collective experience of many spatial data system implementations, software and hardware expenses are typically much less than 20 percent of the total costs. If this is true, then the annual expenditures for spatial data collection, manipulation, and the institutions involved are on the order of $8 to $10 billion, and may be significantly more.

ISSUES

The question addressed by this study is: What could be done better or more efficiently if the content, accuracy, organization, and control of spatial data were different? In reviewing the spatial data activities of a variety of federal agencies, the MSC identified several general issues and impediments that need to be resolved to build a more robust NSDI. Although most issues focus on federal activities, they are parallel to those that exist in state and local governments and the private sector. These issues are discussed in greater depth in Chapter 4.

ISSUE 1: There is no agreed-upon national vision of the NSDI nor is there an apparatus to implement it. Consequently, there is no national policy covering spatial data nor is there a national organization or agency with the charter, authority, and vision to provide leadership of the nation's spatial data collection, use, and exchange.

ISSUE 2: Because of the lack of central oversight, there appears to be extensive overlap and duplication in spatial data collection at the federal level. Overlap in data collection also appears to occur between federal and state agencies, and among state, local, and private sector organizations, all at a significant cost to the public. These institutions are collecting spatial data at many scales, levels of accuracy, levels of detail, and categories of data, making the sharing of spatial data very difficult (if not impossible).

ISSUE 3: There are no current mechanisms that allow identification of what spatial data have been collected, where the data are stored, who controls the access to the data, the content of the data, and the data coverage (e.g., scale, data density).

ISSUE 4: Although a Federal Information Processing Standard (FIPS) for spatial data transfer has been approved, profiles for implementing this standard for the exchange of spatial data between federal agencies have yet to be developed. Moreover, standard activities need to be expanded beyond transfer standards to include more specific measures and standards of content, quality, currency, and performance of various components of the NSDI. As a corollary, there is no agreed-upon representation of "base data" for small-, medium-, and large-scale spatial data products.

ISSUE 5: There are major impediments to, and few workable incentives for, the sharing of spatial data among the federal, state, and local organizations.

The committee studied two broad areas of intense spatial data activity: urban fabric (Chapter 5) and wetlands (Chapter 6 and Appendix B). In particular, our study of the relationship of wetlands and the spatial data that describe them yielded a myriad of problems that might be helpful in addressing similar complex environmental issues of national interest, for example, the geographic distribution of endangered species or the monitoring of bio-diverse lands. The pervasiveness of these issues is apparent in both the urban fabric and wetlands examples.

RECOMMENDATIONS

In response to the above issues and others discussed within in the report, we offer a series of recommendations intended to strengthen the NSDI and make it more robust. That is not to say that additional efforts should not be pursued and encouraged. These recommendations are discussed in detail in Chapter 9.

1. **Effective national policies, strategies, and organizational structures need to be established at the federal level for the integration of national spatial data collection, use, and distribution.**

2. **The Federal Geographic Data Committee (FGDC), which operates under the aegis of the Office of Management and Budget (OMB), should continue to be the working body of the agencies to coordinate the interagency program as defined in OMB Circular A-16. However, the charter and programs of the FGDC need to be strengthened to**

- **expand the development and speed the creation and implementation of standards (content, quality, performance, and exchange), procedures, and specifications for spatially referenced digital data, and**
- **create a series of incentives, particularly among federal agencies, that would maximize the sharing of spatial data and minimize the redundancy of spatial data collection.**

3. **Procedures should be established to foster ready access to information describing spatial data available within government and the**

private sector through existing networks, thereby providing on-line access by the public in the form of directories and catalogs.

4. **A spatial data sharing program should be established to enrich national spatial data coverage, minimize redundant data collection at all levels, and create new opportunities for the use of spatial data throughout the nation. Specific funding and budgetary cross-cutting responsibilities of federal agencies should be identified by the OMB and the FGDC should coordinate the cross-cutting aspects of the program.**

CONCLUSION

This country has a tradition of localized control in the public sector and a belief in the power of free market forces operating in the private sector to best serve the national interest. In an era of instantaneous nationwide and worldwide transmission of information, compartmentalization of spatial data collection and management may no longer make sense as it has in the past. *Survival in an increasingly global economy, dominated by ever larger private-public sector coalitions in countries outside the United States, may be possible only if commitments are made in this country to a national policy for increased information development and sharing.*

2
INTRODUCTION

Before about 1960 the dominant medium for recording and transmitting information about geographic location was the map, a highly conventionalized analog image, usually ink on paper. The map has certain unique values as well as profound limitations.

In a sense, the total store of information about the geography of a nation in the past was equivalent to all existing maps. Because it is difficult or impossible to aggregate or cumulate maps or mapped information in any practical way, this store of information was widely dispersed and frequently encapsulated in particular applications. Except for a few major programs, most importantly the national topographic map series produced by the National Mapping Division (NMD) of the United States Geological Survey (USGS), mapping has traditionally been application specific. Maps are almost never made strictly for the sake of mapping; they are tools, having value only in allowing their users to do other tasks: for example, manage land records, run paper companies, build highways, carry out a census, or search for minerals.

Many public and private sector organizations that traditionally made maps or caused them to be made were not primarily mappers and viewed mapping strictly as a cost of doing business. Once used or applied, much map information was stored or discarded. It never became accessible to other potential users, and thus could not be said to contribute significantly to a national store of knowledge.

One could in the pre-1960 era conceive of a social-economic entity called "the cartographic enterprise," that is, all individuals and institutions involved in the production, use, and dissemination of maps. Except for the widespread use of USGS topographic maps as a source of common base data, there was little integration among the components of this entity. The cost and time associated with the compilation, drafting, printing, and distribution of paper maps prohibited such integration.

The introduction of the digital computer in the 1960s and 1970s led to two major developments in mapping. First, the printed map can be produced by using the new, more flexible digital technology. Second, the printed map is increasingly being supplemented and replaced by computer-based geographic information systems (GIS), which treat maps as a series of spatially integrated layers. While paper maps originated to supplement human memory and vision, GIS do this and more: they supplement human cognitive or information-processing capabilities as well.

In the GIS era it is no longer adequate to speak of a "cartographic enterprise" with its connotation of simply making paper maps. Instead we must develop new conceptions more useful in describing and analyzing how geographic data (any data referenced to location) are acquired, processed, disseminated, and used.

Unlike maps, strings of geographic or spatially referenced digital data can be aggregated, transformed, and shared. Spatial data can now be more easily isolated and abstracted from the particular application in which it was developed and channelled into other settings and other GIS where it can be reused, enhanced, and routed to other potential user communities. The old "top down" model (especially appropriate for base data from NMD and other federal agencies) is inadequate to represent the multidirectional alternative information flows that are now technically feasible.

Spatially referenced digital data can perhaps be thought of as molecules of water that in aggregate form a circulating fluid, flowing freely from application to application (or from system to system). Conceived of in this fashion, information can be seen to take on value and become a marketable commodity, quite apart from the context, need, or application for which it was originally developed.

An important qualification to this concept must, however, be clearly stated. It is often said that our nation's economy is becoming information-based, and statistics are produced to show its considerable economic value. It is critical to recall (as many journalists do not) that while information seems to have a reality of its own, it takes on value *only* with reference to authentic value-producing activity, that is, only when it is about something. For example, even though a timber products firm might like to buy rather than produce basic mapping of its resources and facilities and might assume that there would be a market for such information *per se*, the need and therefore the market for such data may exist only as long as the primary business (timber production and sale) itself exists. This is obvious but often seems to be overlooked in discussions of the information economy.

Considerable quantities of spatial data are generated for special or even unique purposes, and it is unlikely that much of it needs to circulate widely. In addition, proprietary business data, conceivably of very general interest, also would not be publically shared.

Because of rapidly advancing capabilities and uses of technology, it is often difficult to understand and visualize the most profound implications of the computer for handling geographic information. GIS require vast amounts of digital spatial data to function and because these data are not commonly derived by digital sensing of geographic reality (much of which is not physical anyway), the users have been forced to rely on data derived from analog maps. Consequently, much of the total investment in GIS for decades has been in a continuing, tedious, and expensive process of converting analog (paper) maps into digital data bases. Furthermore, given the application-oriented nature of mapping, including GIS, the committee has seen examples of waste and redundancy as the same maps are repeatedly digitized by different organizations.

The maps or categories of information on maps that are repeatedly being digitized at high cost (primarily to the tax-paying public) are what are commonly called base maps or base data. It can be safely assumed that the most commonly digitized base maps are the 55,000 sheets of the NMD's 1:24,000-scale topographic map series and that the most commonly digitized data categories include graticule, shorelines, drainage, political boundaries, and transportation routes.

In retrospect, it might have been wise early on to digitize these categories nationwide, as a one-time-only crash effort. This was not done, and the nation continues with a disproportionate share of its spatial data resources devoted to repeating base data digitizing rather than creating new, improved data.

The MSC (1990), in its report *Spatial Data Needs: The Future of the National Mapping Program*, focused on the role played by the USGS in providing spatial information to other federal agencies and to the nation as a whole. During most of the USGS's history, this information took the form of topographic maps, most importantly and most recently nationwide coverage at the scale of 1:24,000.

However, in the past few decades the demand for spatially referenced information shifted dramatically from analog maps to digital data. As a result, the USGS, the only organization offering a comprehensive national reference map series, came under intense pressure to respond to this change with an altered or expanded program. *Spatial Data Needs* was produced at the USGS's request, and it contained a series of recommendations designed to assist the USGS to adapt to changing user requirements.

In *Spatial Data Needs* the phrase "national spatial data infrastructure" appeared several times, its meaning clarified only by context. At that time no effort was made by the MSC to define the concept precisely, or to examine the roles played by government agencies other than the USGS. However, in 1990 the responsibilities of the MSC changed to include a broader perspective on federal mapping organizations (with the word mapping used here as a very general term encompassing all spatial data collection and use) and, less directly but crucially, the mapping activities of the nation as a whole. With this change in perspective, the need to better articulate our objectives became apparent. Therefore, the MSC considers that its mission includes responsibility for the NSDI as follows:

> *The Mapping Science Committee will serve as a focus for external advice to the federal agencies on scientific and technical matters related to spatial data handling and analysis. The purpose of the committee is to provide advice on the development of a robust national spatial data infrastructure for making informed decisions at all levels of government and throughout society in general.*

The MSC has for 2 years been gathering information about the programs of a wide variety of mapping organizations in the public sector (federal, state, and local) as well as from a more limited selection of firms in the private sector. The MSC is now in a position to provide both a more comprehensive definition of a national spatial data infrastructure and a series of recommendations to improve its utility. The purpose of this report is to:

- describe the concept and evolution of the NSDI;
- document the success or failure of the current NSDI, with evidence based on an examination of specific spatial data domains;

- analyze impediments to success and, where possible, propose ways these impediments can be removed; and finally,
- propose the creation of incentives, which probably will nee to involve many institutional changes, that will lead to the creation of a more useful and cost-effective NSDI for the nation as a whole.

REFERENCE

MSC (1990). *Spatial Data Needs: The Future of the National Mapping Program*, Mapping Science Committee, National Research Council, National Academy Press, Washington, D.C., 78 pp.

3
NATIONAL SPATIAL DATA INFRASTRUCTURE

EVOLUTION

In the past few decades, there has been an evolution of how data and information are perceived in the context of an information society and in the capabilities of handling spatial data. The changes have been remarkably parallel.

The characteristics of data and information have progressed from a discipline orientation to a mission orientation and finally to a problem orientation (American Library Association, 1978). All three orientations currently exist and are interrelated. The disciplinary orientation largely concerns physical measurements often of a narrow technical nature (e.g., hypsography). The mission orientation represents the organization of data and information required for a specific job (e.g., building a weapon system or assessing a natural resource). The problem orientation draws on data and information from the disciplinary and mission orientations but is distinguished by the need to integrate information from a variety of sources primarily for public policy formulation and decision making. The characteristics of these three orientations are given in Table 3.1.

In a similar context, efforts to build computer-based spatial information systems can also be viewed in three stages. The initial stage dates back to the early 1960s, evolved through a second, and is now entering a third (McLaughlin, 1991). There is no simple boundary separating stages. The dates given below are somewhat arbitrary but they represent major shifts in the general environment.

TABLE 3.1 Characteristics of Data and Information Orientations

Orientation	Information Content	Target User Group	User Sophistication
Discipline Oriented	"Truth of science," mostly dealing with physical systems	• Members of the discipline involved • Generally individuals heavily committed to disciplines	Very high technical sophistication; communicates to a particular technical elite
Mission Oriented	The above; also truth of science as it applies to technological objectives, ways in which physical systems can be manipulated and controlled for practical purposes	• Technologists, engineers, and applications people • Team members committed for multiyear periods	Moderate to high technical sophistication; needs to communicate across disciplines
Problem Oriented	All of the above; also wisdom and pragmatics that work in contexts of uncontrollable open systems involving human as well as physical dynamics; includes legal, sociological, economic, and demographic information and important social value inputs	• Large segments of the public in state and local governments and all types of industries, communities, planners, conservationists, bureaucrats, and policy makers at all levels • Transient, relatively short-term users	Very low to high technical sophistication; needs to communicate across major cultural groups; heavy involvement of nontechnical professionals and political groups

Source: American Library Association (1978)

Stage I (circa 1960-1980)

During this first period there was much experimentation and a genuine effort to develop a new paradigm for managing spatial data. With its focus on research and development, this period saw the first use of computers in surveying and mapping, the first efforts in automating land records, and the first attempts to build urban and regional information systems. The inspiration largely was driven by the public value (or need) of spatial data rather than the technology or commercial value of the data.

There was a broad vision of an information society, including the inherent power of spatial data systems to help resolve land issues and to enable greater citizen participation in decision making. In government and university laboratories, the first GIS software was developed to integrate the variety of information required for regulation and land-use planning. State systems in Minnesota and elsewhere evolved in response to the environmental agendas of the late 1960s and early 1970s. Furthermore, new concepts, such as the multipurpose cadastre, emerged for organizing land-related activities around shared information. The vision cut across disciplinary boundaries and the early years were characterized by a strong interest in integration and cooperation.

Stage II (1975-2000)

If the first stage was marked by innovation and a broad vision, then the second stage might be described as a retreat. This is the period in which computer-based spatial information systems have come into their own for administration, facilities management, and planning. Governments at all levels are making large investments in data base development. But for the most part these accomplishments are occurring within traditional institutional models (e.g., line departments automating their established activities). Far too often, the vision of the integrating powers of the new technology and the potential devolution of power to the people has been lost.

A critical feature of this period has been the rapid emergence of commercial GIS and digital mapping software, and by the mid-1980s a major vendor-driven market had been established. The vendors increasingly set the agenda. The tools of the trade began to dominate the proliferation of conferences, and the question of why was replaced with how. No organization has wanted to be left out but few have had innovative ideas about using the tools they purchased. Despite emphasis on graphics and spatial analysis (especially by academics), the predominant interest has been in textual data and in simple, straightforward query applications of the spatial data. There have been important technological improvements (especially in hardware) but many of the basic concepts date back to the first stage of research and development.

Stage III (1990 and beyond)

Now that there have been major investments in technology and the process of building data bases has begun, there is an opportunity to regain a sense of purpose. Infrastructure is different from information systems: the emphasis is shifting once again from information technology to information and its use in society. As the technology becomes commonplace (shown by the use of terms such as appliance), the emphasis on the simple query tools will diminish. Information will become part of the background as communica-ting and extracting knowledge become increasingly important.

The period we are gradually moving into will be dominated by integration and applications. The focus will be on linking data bases into distributed spatial information networks and on developing application software and decision-support tools to more effectively exploit the available information. What has been perceived as having commercial value in the information industry will change with the diffusion of value-added products and services. Social issues such as rights of privacy and access, which have mostly been given lip service, will require real solutions. New institutional arrangements, alluded to in Stage I and largely dismissed in Stage II, will finally begin to emerge. Driving this movement will be a renewed concern about the effective development and management of our land and marine resources and the relation to the larger environmental agenda.

CONCEPT

We are beginning to see coordination, linkages, and information "flows" that are made possible by the application-neutral character of digital spatial data and by the power of the computers and networks used to process and distribute the data. There has been a technological revolution, but the institutional evolution required for us as a nation to take advantage of it has not yet occurred. The balance of this report will examine this discrepancy in detail to effect what we will argue is much-needed change.

In summary, *the National Spatial Data Infrastructure is the means to assemble geographic information that describes the arrangement and attributes of features and phenomena on the Earth. The infrastructure includes the materials, technology, and people necessary to acquire, process, store, and distribute such information to meet a wide variety of needs.*

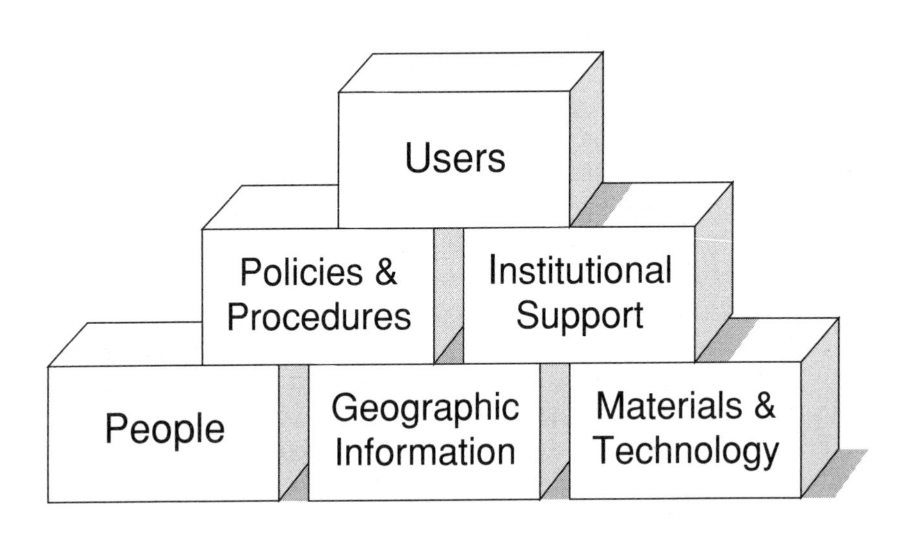

Figure 3.1 NSDI building blocks.

Although spatial data are necessary for a spatial data infrastructure, they are *not* sufficient. Of equal importance are the individuals, institutions, and technological and value systems that make it a functional entity, one that serves as a basis for much of the business of a nation.

It should also be made clear that the word national, as used in NSDI, does not refer exclusively either to the federal government or to data coverages for the whole country. The MSC uses the term as applicable to the needs of the nation as a whole, in the interests of the public and private sector organizations, and individuals nationwide who provide the financial resources to support it.

Finally, we must emphasize that there *is* a national spatial data infrastructure in existence. It is an *ad hoc* affair because, until very recently, no one has conceived of it or defined it as a coherent entity, and indeed it has not been very coherent or coordinated. It is not the task of the MSC (or of anyone else for that matter) to *create* a national spatial data infrastructure. We want merely to point out its existence, identify its components and characteristics, assess the efficiency and effectiveness with which it functions to meet national needs (particularly at the federal level), and make recommendations that might make it more useful, more economical, more effective, better-coordinated, and robust. Several investigators have shown that investments in using GIS are normally repaid by the long-term benefits and that greater efficiencies are realized (Brown, 1990; Gillespie, 1992). In addition, there could be a significant reduction in the cost of operation of GIS if existing data were shared, thus reducing duplication of efforts of data collection.

Information flow, particularly in the spatial data infrastructure, should be seen as critical to the maintenance of a competitive position for the United States in an increasingly international economic arena. The concept of a spatial information infrastructure has recently been addressed in a number of countries, including Australia, Canada, New Zealand, and the United Kingdom (Newton *et al.*, 1992; Rhind, 1992).

Once the national spatial data infrastructure is recognized, its functions and malfunctions can be examined in detail. The MSC found two major categories of obstacles to efficiency and change. First, there are *serious impediments* to coordination and efficiency. Traditional arrangements for creating, disseminating, using, and paying for maps do not necessarily work in the spatial data era. Consequently, failure to recognize information creation and distribution as legitimate parts of the mission of federal agencies often creates enormous waste for the nation's taxpayers.

Second, there is no *system of incentives* to direct funding and capability, both public and private sector, in the direction of more coordination and greater efficiency. The creation of incentive systems will, among other things, make it attractive and perhaps even essential for the public and private sectors to work together in the spatial data infrastructure to a degree undreamed of in the past.

REFERENCES

American Library Association (1978). *Into the Information Age: A Perspective for Federal Action on Information*, American Library Association, Chicago, 134 pp.

Brown, K. (1990). *Local Government Benefits from GIS*, PlanGraphics, Inc., Lexington, Kentucky.

Gillespie, S. R. (1992). The value of GIS to the federal government, in *Proceedings: GIS/LIS '92 Annual Conference and Exposition*, Anaheim, California, pp. 256-264.

McLaughlin, J. (1991). Towards a national spatial data infrastructure, in *Proceedings of the Canadian Conference on GIS*, Ottawa, March 1991, pp. 1-25.

Newton, Zwart, and Cavill, eds. (1992). *Networking Spatial Information Systems*, Belhaven Press, London.

Rhind, D. (1992). The information infrastructure of GIS, in *Proceedings, 5th International Symposium on Spatial Data Handling*, International Geographical Union Commission on GIS, August 3-7, 1992, Charleston, South Carolina, pp. 1-19.

4
CURRENT SITUATION

It is important to conceptualize the NSDI in the broad sense of our definition. However, when the committee received briefings from various agencies and studied the documentation provided, an important area of opportunity concerned data. The issues concerning the reange of spatial data are complex and hence we offer no panaceas, but we have attempted to discuss some of the impediments and improvements to using and sharing data in the NSDI.

The data in the NSDI exist in diverse forms and reside in the analog and digital spatial data bases of various federal, state, local, and private agencies. Parts of these data exist as hard copy maps and charts created and periodically updated to meet a defined need. Other parts are stored as reports and studies that summarize the spatial data standards, needs, uses, and controls of the federal, state, local, and private sectors. Improvements can be made in the underlying structure, procedures, and standards that would allow for easy exchange of these data.

It is important to note that data sharing has occurred. When needed data were found to be unavailable, an organization or agency would set about collecting them. If all or part of the needed data were available from another agency, a data exchange was often set up. When agencies found they were responsible for data collection over the same areas, coproducer agreements were set up where feasible. However, most data collection overlaps were resolved by each agency collecting its own data over the same area, ostensibly to meet needed accuracy and currency requirements. The justifications for this duplication in data collection usually fell into three categories: (1) it would be too expensive to collect the level of data needed by one agency to satisfy the needs of a second; (2) the accuracy or level of aggregation required by the first agency could not be achieved by the data collection capabilities of the second agency; and (3) the data could not be collected in the time frame required by the other agency.

With the advent of GIS, an extremely versatile, efficient tool has become available for manipulating spatial data. The diverse types of GIS applications have driven the need for more timely and accurate spatial data, particularly in digital form. This also has created the need data base products that are adapatable to rapidly changing user needs.

As federal agencies discovered the value of using GIS for storing and manipulating spatial data, budgets were constructed to purchase such equipment and to collect the needed data in digital form. Early efforts resulted in a myriad of data formats, standards, and processing algorithms. In effect, each federal agency repeated what it had done with paper products, only in digital form.

The cost (in staff hours) of collecting spatial data in digital form is projected to be half the cost of collecting data by manual methods for the generation of individual products (DMA, 1991). Once collected, though, the value of such digital data increases manyfold. The ability to extract subsets, generalize, and thin the data; increase its densification; or merge (fuse) it with other spatial data has created such exciting applications as emergency response (911) location systems, integrated land use/transportation planning models, battlefield management systems, crop rotation forecasts, and environmental impact assessments.

MAJOR ISSUES

While reviewing the spatial data activities of several federal agencies, the MSC recognized a number of general issues and impediments that need to be resolved to build a more robust NSDI.

ISSUE 1: There is no agreed-upon national vision of the NSDI nor is there an apparatus to implement it. Consequently, there is no national policy covering spatial data nor is there a national organization or agency with the charter, authority, and vision to provide leadership of the nation's spatial data collection, use, and exchange.

Each federal agency with a responsibility for collecting spatial data traces this responsibility to the fulfillment of its primary mission. For example, the spatial data collected by the USGS satisfy its requirement to produce specific products from the collected data. Agencies must focus their funding toward satisfying their primary mission and mandates, and few, if any, resources are left to support government-wide spatial data oversight activities. In fact, the current cooperative efforts between agencies are more a result of bilateral agreements for data exchange than a concern for a healthy NSDI.

The Federal Geographic Data Committee (FGDC) was established by revised OMB Circular A-16 as an interagency coordinating committee on geographic data matters. However, the present direction and organization of the FGDC have problems that inhibit its effectiveness. The FGDC has no charter to review the spatial data programs of its members and no power to enforce decisions. Because there is always some resistance to change, even FGDC recommendations on spatial data content and format take time to implement. The federal agency members of the FGDC steering committee have varied interest and involvement in collecting and applying spatial data. Some, in fact, seem to have a vested interest in maintaining the status quo. Finally, the agency representatives are not detailed to the FGDC as their primary mission, and their job commitments remain with their representative agencies, not with the FGDC.

ISSUE 2: Because of the lack of central oversight, there appears to be extensive overlap and duplication in spatial data collection at the federal level. Overlap in data collection also appears to occur between federal and state agencies, and among state, local, and private sector organizations, all at a significant cost to the public. These institutions are collecting spatial data at many scales, levels of accuracy, levels of detail, and categories of data, making the sharing of spatial data very difficult (if not impossible).

As shown in Table 4.1, many federal agencies are responsible for collecting spatial data of the same type over the same areas of the country. In the area of wetlands data collection (see Chapter 6 and Appendix A), for example, overlap of data collection is significant.

Additional redundancies in data collection result from the diversity of definitions used for various classes of data. For instance, the Soil Conservation Service (SCS) and the Agricultural Stabilization and Conservation Service (ASCS), using the National Resource Inventory (NRI), collect resource data on a sample basis for all non-federal lands. The Bureau of Land Management (BLM) and the U.S. Forest Service (USFS) collect similar data for the land under their jurisdictions. The Environmen-

Table 4.1 Examples of the Range of Spatial Data Collection Responsibilities

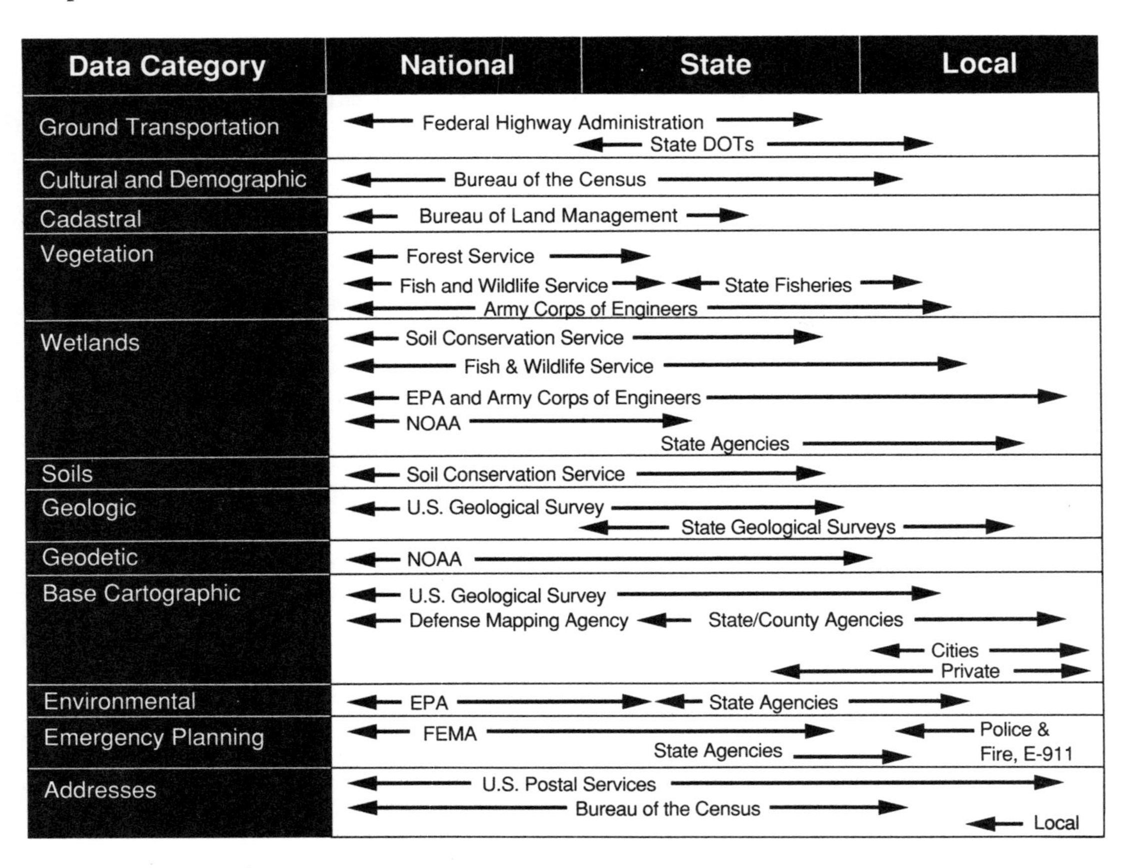

tal Protection Agency (EPA) has proposed to collect a nationwide sample of environmental conditions. All these agencies and programs use different primary sampling units and slightly different types of data. Data delineating wetlands are being collected by the Fish and Wildlife Service (FWS), the EPA, the SCS, the USFS, the National Oceanic and Atmospheric Administration (NOAA), the USGS, and many state agencies. However, there is still no uniform approach to the definitions and mapping conventions among these programs.

As mentioned, the primary data collection role of most federal agencies is to supply data products to meet their own missions and mandates. These agencies create products of differing scale and content and use different collection techniques and source materials. Often, such data are incompatible with similar data collected for other products without algorithms to thin, match, feather, generalize, densify, remensurate, rescale, or aggregate data categories. This problem is further complicated by the variances in the underlying base data, non-standard feature identifiers, and GIS hardware and software used to store and manipulate the data.

ISSUE 3: There are no current mechanisms that allow identification of what spatial data have been collected, where the data are stored, who controls the access to the data, the content of the data, and the data coverage (e.g., scale, data density).

Each federal agency controls its own spatial data as do state and local governments and private sector organizations. Some agencies, such as the USGS and the Bureau of the Census, have well-publicized mechanisms for obtaining their data products. Others, such as the EPA, make their data available to qualified users through an on-line system. Some agencies have yet to institute any type of formal mechanism for distributing their data. These data often can be obtained only by dealing with the person who controls the data within a specific agency. In summary, there is no single system that provides a catalog or directory of the spatial data holdings of all federal agencies. In January 1993, the Federal Geographic Data Committee printed a *Manual of Federal Geographic Data Products*, which represents a positive first step in the compiling the various federal spatial data holdings.

Even if a comprehensive spatial data catalog could be developed for the federal agencies, the extension of such a system to state and local government holdings would be very difficult because to its potential size.

Extension to the private sector might further complicate the process by introducing proprietary restrictions.

There is technology available to provide a distributed, on-line system that provides access to spatial data catalogs at various organizations. One such system is a public-domain software program, *Wide Area Information Servers* (WAIS). WAIS, which runs on the Internet network, provides the capability to scan, search, and often access existing data bases. WAIS could be easily expanded to encompass spatial data catalogs. (See Chapter 8 for additional information on WAIS and spatial data catalogs.)

ISSUE 4: Although a Federal Information Processing Standard (FIPS) for spatial data transfer has been approved, profiles for implementing this standard for the exchange of spatial data between federal agencies have yet to be developed. Moreover, standard activities need to be expanded beyond transfer standards to include more specific measures and standards of content, quality, currency, and performance of various components of the NSDI. As a corollary, there is no agreed-upon representation of "base data" for small-, medium-, and large-scale spatial data products.

There is general agreement that spatial data exchange standards are a federal responsibility. However, there is a plethora of standards already identified for the exchange of spatial data. Most of these concern special data exchanges between a data producer and its user community or between GIS software systems. Other exchange standards are in the form of specifically defined products such as TIGER/Line, Digital Chart of the World (DCW), Digital Terrain Elevation Data (DTED), Digital Feature Analysis Data (DFAD), or Digital Line Graphs (DLGs). These files of specific spatial data have a particular format that the sending and receiving organizations have agreed to use for data exchange. These product formats, however, are not robust enough to support the ad hoc exchange of digital data.

The Spatial Data Transfer Standard (SDTS; FIPS-173) is an attempt by federal agencies, under the sponsorship of the FGDC, to develop a general family of exchange standards for civilian geographic data. The SDTS should serve as the umbrella for more specific "federal profile" standards, such as DIGEST, VPF, and TIGER.

With regard to base data, they should include, at a minimum, those spatial and primary attribute data that can identify all relevant information for a particular scale product. The base data must meet relevant positional accuracy and unique identifier requirements for inclusion of value-added data from other organizations. Table 4.2 includes a typical set of base data that are representative for three scale ranges.

Instead of a single base of data, the agencies are using multiple bases that are often incompatible. Some agencies are using the Bureau of the Census' TIGER/Line files as their base, whereas others are using the USGS 1:24,000 quadrangles. Although TIGER contains street names, the DLGs do not. Therefore, it is difficult to transfer TIGER attributes to the more geographically accurate DLGs. Other agencies such as the EPA and the SCS's NRI are using unique sampling units on separate base data at varying scales. Some data collection efforts such as the proposed digital orthophoto program (SCS, ASCS, and USGS) and the Digital Chart of the World [produced by the Defense Mapping Agency (DMA)] collect data to still different base scales and accuracies. These two products could become the base data standards of reference in the future for large- and small-scale products that would complement the USGS 1:24,000 (medium scale) DLG products, which have become the de facto medium-scale standard.

ISSUE 5: There are major impediments to, and few workable incentives for, the sharing of spatial data among the federal, state, and local organizations.

All federal agencies recognize the large expense involved in collecting spatial data in digital form. However, there are no real incentives (in fact there are oftentimes disincentives) to the sharing of these data. To share data, the federal agencies would have to agree in advance on the uses, types, and formats of the collected data. This would require one agency to collect, at its expense, more data than it required to satisfy another agency's needs. At the same time, the agency would become dependent on other agencies to collect data with the needed rigor, accuracy, and urgency. If such agreements could be arranged, one or both agencies might lose personnel and funding as a result of the savings in spatial data collection.

Collection of spatial data for sale to the general public has its own set of disincentives. If state, local, or private organizations are willing to pay for the data, the payments often go to the general U.S. Treasury and are not returned to the agency that dedicated resources to collect and distribute the data to the buyer. The USGS is one of the few agencies that are allowed to collect proceeds from map and data sales and directly reimburse the relevant programs. Without a way to be directly reimbursed for the

TABLE 4.2. Examples of Base Data

	LEVEL I	LEVEL II	LEVEL III	LEVEL IV
ACCURACY				
Range	1:500 to 1;10,000 map; Class A NMAS	1:10,000 to 1:100,000 map; Class A NMAS	1:100,000 to 1:500,000 map; Class A NMAS	1:500,000 and smaller; Digital Chart of the World (DCW)
Nominal scale	1:5,000 — Center line with feature dimension	1:24,000 — Center line with feature dimension	1:250,000 — Center line with feature dimension	1:1,000,000 — DCW
CONTENT				
Elevation	0.5 to 2 meters	2 to 20 meters	20 to 40 meters	——
Orthophoto	0.2 to 2.0 meter GRD	1 to 10 meter GRD	10 to 70 meter GRD	——
Transportation[a]	As shown on 1:5,000	As shown on 1:24,000	As shown on 1:250,000	DCW Specifications
Other culture	As shown on 1:5,000	As shown on 1:24,000	As shown on 1:250,000	DCW Specifications
Soils	As shown on 1:5,000	As shown on 1:24,000	As shown on 1:250,000	DCW Specifications
Vegetation	As shown on 1:5,000	As shown on 1:24,000	As shown on 1:250,000	DCW Specifications
Names	As shown on 1:5,000	As shown on 1:24,000	As shown on 1:250,000	DCW Specifications
Political Boundaries	As shown on 1:5,000	As shown on 1:24,000	As shown on 1:250,000	DCW Specifications
Hydrology	As shown on 1:5,000	As shown on 1:24,000	As shown on 1:250,000	DCW Specifications
Topology	Yes	Yes	Yes	Yes
PLSS	Yes	Yes	No	No
Geodetic control[b]	All first and second order control points, with descriptions	Spatially distributed representative points	Spatially distributed representative points	Spatially distributed representative points

[a] Transportation serves as a geometric supplementary framework.
[b] Geodetic control is included to the highest known accuracy.

resources spent, federal agencies are reluctant to collect spatial data for others.

Even when joint memoranda of agreements are signed between federal agencies, it is usually for equal exchanges of effort. Data collected by one agency in one part of the United States are given to a second in exchange for equivalent data collected elsewhere. Examples of this type of joint data collection include the proposed USGS/ASCS/SCS orthophoto program and the use of USGS 1:24,000 quadrangles by many agencies as the base for their products. Other examples, such as the seemingly redundant resource mapping done by the SCS, the National Agricultural Statistical Service (NASS), the USFS, and the EPA's proposed Environmental Monitoring and Assessment Program (EMAP), show a distinct proclivity for not sharing data, sometimes even within the same department. At a minimum, coordination of sampling activities should occur within and between federal agencies.

FEDERAL AGENCY ACTIVITIES

The major issues reflect problems that the federal agencies are experiencing with the use of spatial data. Although customer demands for spatial data have increased dramatically and GIS technology has kept pace, the federal "suppliers" of spatial data have had significant difficulty converting their hardcopy production process to flexible production systems capable of supplying spatial data in both digital and printed form.

A review of current federal agency activities (see Table 4.3) illustrates the magnitude of these problems. In general, each agency is trying to overcome the problems within the confines of its own charter, thus erecting additional impediments to establishing a robust NSDI. Most federal agencies are also concentrating on supplying spatial data in forms that match defined hardcopy products rather than in forms that lend themselves to easy manipulation by commercial GIS packages. In this regard, the federal agencies have not had incentives to meet their evolving mission as suppliers of spatial data to their customers.

The MSC paid special attention to federal agencies with major spatial data collection efforts: the Bureau of the Census, the BLM, the DMA, the USFS, the FWS, the SCS, and the USGS. The EPA is discussed because it is one of the largest federal users of spatial data. These agencies will be

TABLE 4.3 Examples of Federal Use of Geographic Information

Federal Organization	Selected Activities Making Use of Geographic Data
Department of Agriculture	Plant and animal disease quarantine studies; pest management; cropland, forest, and rangeland management; soils mapping; watershed quality, planning, and management
Department of Commerce	Census taking; climatic mapping; definition, establishment, and maintenance of the National Geodetic Reference System; development of geoid models; global modeling; coastal, estuarine, and marine resource monitoring and management
Department of Defense	Base master planning; facilities siting and management; resource and land use management; environmental analysis and planning; hazardous waste remediation; mapping and charting; tactical and strategic military operations
Department of Energy	Environmental impact analysis; facilities siting; transmission line routing; hazardous waste remediation; energy resource estimates; emergency response; evacuation planning; hazardous materials routing
Department of the Interior	Natural and cultural resource management and planning; economic development; transportation planning; alternative and conflicting use analysis for federal lands; mineral and energy resource analysis; mineral, oil, and gas leasing; water rights issues; environmental monitoring; mapping; wetlands inventory and trend studies; cadastral surveys; habitat suitability analysis; land records management; water quality evaluation
Department of Justice	Litigation; voting rights protection; drug enforcement
Department of Transportation	Airspace management; economic impact of highways studies; intelligent vehicle and highway systems; policy analysis; emergency response planning; traffic engineering; vehicle routing
Environmental Protection Agency	Air quality monitoring; Superfund site discovery analysis; Resource Conservation and Recovery Act site management; water quality and vulnerability studies; risk assessment
Federal Emergency Management Agency	Emergency planning and response; flood insurance program administration
U.S. Postal Service	Mail collection and delivery route modeling
Tennessee Valley Authority	Land and reservoir management; facilities site screening; natural resource and economic development; facilities management

Sources: FICCDC (1990); FGDC (1992).

discussed in the light of their responsibilities to meet OMB revised Circular A-16.

Office of Management and Budget

The OMB has executive oversight responsibility for the budgets of all federal agencies and for establishing policies and procedures for federal agency GIS activities. Part of this oversight includes the review and approval of all large expenditures for spatial data systems. The OMB issued Circular A-16 in 1953 (revised in 1967 and, most recently, in 1990) to facilitate coordination of federal mapping activities. In 1983, an OMB memorandum established the Federal Interagency Coordinating Committee on Digital Cartography (FICCDC) to assist in coordinating data sharing across federal agencies. The FICCDC found many obstacles that prevented data sharing between government agencies and the private sector; barriers were of a technical, institutional, and legal nature. The FICCDC cooperated in developing and publishing the first draft of the SDTS to help mitigate these problems.

One recent concern of the OMB has been the shifting of the nation's map base to electronic media. In 1988, the OMB recognized the need to look for opportunities to share data, standards, and development efforts among the federal agencies. Such sharing should substantially reduce overall costs to the U.S. government.

However, the expected costs continue to rise for developing, implementing, and operating systems to convert existing map graphics from paper or other stable base to electronic form. The results of the OMB Bulletin 88-11 survey in 1988 showed that the estimated costs for "electronic mapping" of the civilian side of the federal government exceeded $100 million for FY1988 and would top $200 million by FY1992. (Briefings by the federal agencies to the MSC indicate that the estimates for current federal mapping initiatives are much greater—by at least a factor of two—than indicated by the 1988 OMB Bulletin.)

The civilian agencies with the largest expenditures in spatial data activities are the EPA, the Bureau of the Census, the BLM, the SCS, the USFS, and the USGS. The EPA indicated that their FY1992 investment in spatial data was around $500 million; the SCS reported that their expenditures were around $220 million. (Both the EPA and the SCS alone exceed the 1988 OMB estimate; however, caution needs to be applied to what is included in these estimates.) In other agencies the MSC found that modernization efforts accounted for 75 percent of their estimated mapping-related expenditures.

OMB is especially interested in cost savings and return on investment. In particular, new development efforts must show short-term, quantifiable payoffs from investments. Most important are Class I savings, which show direct payoffs for the investment. Cost avoidance approaches (Class II) are also good, but not as meaningful as the direct savings of Class I. Class III savings, the improvement of goods and services for the public sector, are also saleable. [An example of a Class III savings would be the new Federal Aviation Administration (FAA) system that would make flying safer.]

Applying cost savings approaches for spatial data initiatives includes (1) use of standards for storage and transmission of spatial data (such as the SDTS), (2) maximizing the use of existing data bases and encoding schemes, (3) developing excellent benefit-cost ratios for any new GIS or spatial data initiative, and (4) sharing of spatial data collection and new technology developments among federal agencies.

OMB Circular A-16 was revised in 1990 to encourage data sharing among federal agencies. However, no real incentives are in place to ensure that such data sharing occurs.

Federal Geographic Data Committee

The FGDC was established in October 1990 by revised OMB Circular A-16. The FGDC is responsible for promoting and coordinating the development, use, sharing, and dissemination of GIS data throughout the government. The revised circular established the FGDC as replacing the FICCDC. It also empowered the FGDC to promote the coordinated use and development of mapping, surveying, and other spatial data.

OMB Circular A-16 specifically states that the FGDC:

"• supports surveying and mapping activities, aids geographic information use, and assists land managers, technical support organizations, and other users in meeting their program objectives through;

"• promoting the development, maintenance, and management of distributed data systems that are national in scope for surveying, mapping, and related spatial data;

"• encouraging the development and implementation of standards, exchange formats, specifications, procedures, and guidelines;

"• promoting technology development, transfer, and exchange;

"• promoting interaction with other existing federal coordinating activities that have interest in the generation, collection, use, and transfer of spatial data;
"• publishing periodic technical and management articles and reports;
"• performing special studies and providing special reports and briefings to OMB on major initiatives to facilitate understanding of the relationship of spatial data technologies with agency programs; and
"• ensuring that activities related to Circular A-16 support national security, national defense, and emergency preparedness programs."

The revised OMB Circular A-16 assigned responsibilities to numerous agencies for various FGDC roles. The USGS was identified as the lead agency for the FGDC. Ten subcommittees were established, each responsible for coordinating activities for a category of spatial data. There are also three working groups (standards, technology, and state and local liaison) that deal with issues common to all spatial data categories. Table 4.4 shows the federal agencies with lead coordination responsibility for the major data categories.

The FGDC (1991) in its report to the OMB refers to a National Geographic Data System (NGDS) as a system of independently held and maintained federal digital geographic data bases. The NGDS would encompass the National Digital Cartographic Data Base (NDCDB) of the USGS plus other spatial data bases that are national in scope. The NGDS would include traditional cartographic data as well as thematic data, such as soils, wetlands, geology, vegetation, and demographic data. The NGDS would be an important component of the NSDI.

To ensure proper populating and use of the NGDS, standards must also be developed for each spatial data layer. Some standards (coordinate systems, code sets, and geographic names) already exist. Standards for many other segments or layers of the NGDS must be developed. Lineage, positional accuracy, attribute accuracy, logical consistency, and completeness (all parts of data quality) are needed to ensure data integrity.

Other Federal Agencies

A number of federal agencies have extensive, ongoing, spatial data collection programs, as shown in Tables 4.1, 4.3, and 4.4. At least eight of these—the Bureau of the Census, the BLM, the DMA, the EPA, the FWS, the SCS, the USFS, and the USGS—have national data collections

TABLE 4.4 Geographic Data Coordination Responsibilities Assigned by OMB Revised Circular A-16

Geographic Data Category	Lead Agency
Base cartographic	U.S. Geological Survey, Department of the Interior
Bathymetric	National Oceanic and Atmospheric Administration (NOAA), Department of Commerce
Cadastral	Bureau of Land Management, Department of the Interior
Cultural and demographic	Bureau of the Census, Department of Commerce
Geodetic	National Geodetic Survey, NOAA, Department of Commerce
Geologic	U.S. Geological Survey, Department of the Interior
Ground transportation	Federal Highway Administration, Department of Transportation
Portrayal of certain international boundaries	Office of the Geographer, Department of State
Soils	Soil Conservation Service, Department of Agriculture
Vegetation	Forest Service, Department of Agriculture
Wetlands	Fish and Wildlife Service, Department of the Interior

efforts, which suggest the possibility of data sharing with others. Several of the major programs are briefly described below.

Bureau of the Census

The Bureau of the Census has produced over 1 million different map sheets to support its collection and customer use of census data. However, maps and charts are a supporting activity, not their primary product. The

Bureau of the Census traditionally used three major tools for conducting a census: maps (to tell census takers where they are), address reference files, and geographic reference files. The address reference files are used to match census questionnaires to locations, and the geographic reference files serve as a control inventory of over 7 million census statistical zones.

In 1982 the Geography Division of the Bureau of the Census began developing the Topologically Integrated Geographic Encoding and Referencing (TIGER) System to serve as a single data base incorporating the above three tools for the 1990 decennial census. The TIGER files implemented a new data structure to capture every known street and road in the United States, the name (or names) of each, ranges of address numbers, all the railroads and significant hydrographic features, named landmarks, and other named geographic locations. This information was needed to administer and tabulate the 1990 decennial census of population and housing of the United States and will be used in all future censuses.

The TIGER data base is composed of data from three sources. Scanned 1:100,000 maps from the USGS form the base of 98% of the country. GBF/DIME (geographic base file/dual independent map encoding) files from the 1980 census were used for major metropolitan areas. Data digitized by commercial contractors were used to fill in between the 1:100,000 maps and the DIME files. These data were integrated horizontally and vertically and cut on county boundaries.

The current TIGER files contain the latitude and longitude coordinates of more than 42 million linear features (roads, railroads, water features, and landmarks), address ranges, and political and statistical boundaries. Omitted from the TIGER files are contours, public land survey information, rural route address ranges, and accuracy data. That means the TIGER files must be augmented for most other GIS applications.

TIGER files form one of the most important data sources for states, cities, counties, and utility companies. With proper processing, it may be possible to match these files to USGS terrain data and digitized 1:24,000 quad sheets (as they become available) and eventually to 1:12,000 digital orthophotography to provide very accurate data for all GIS systems. Additional aspects of TIGER appear in the Street Centerline section of Chapter 5.

Some private companies sell improved versions of TIGER; others are using TIGER files as the base for special studies. This usually has been done by converting TIGER files into various layers of data for input to a full-function GIS and then adding data attributes as needed. Some attempts to use TIGER files, such as Bell South's 911 emergency address data base system, were not successful and other sources are being developed. The fact that this $300 million investment in TIGER could not meet these requirements is an indication of the possible need for better coordination.

The Bureau of the Census and the USGS have recently undertaken a project to convert TIGER files to the SDTS format. The initial results demonstrate that TIGER content can be expressed in the SDTS format, but it remains to be seen whether the SDTS format is preferable to current TIGER file format for data exchange.

The major problem with TIGER files is keeping the data up to date. The Bureau of the Census' charter includes an update every 10 years but not a continual update cycle. The broad use of TIGER in both the public and private sectors underscores the importance of an up-to-date TIGER system. The Bureau of the Census is investigating the establishment of various maintenance agreements with the states so that TIGER data can be updated more frequently.

Bureau of Land Management

The BLM administers 272 million acres of public lands primarily in the western United States and Alaska. In addition to the surface the BLM is responsible for the subsurface for an additional 300 million acres of public lands in which the federal government has retained the minerals interest (includes lands under Forest Service jurisdiction, wildlife refuges, and military reservations). The basic mission of the BLM includes three primary responsibilities: (1) the public land survey system (PLSS); (2) maintenance of land and minerals records, including title transfer from federal to state and private ownership and all the land records of land that the BLM currently administers; and (3) natural resource management for multiple use. This mission has led the BLM to invest in a variety of land and geographic information systems. The BLM has been designated the lead agency for the cadastral subcommittee of the FGDC.

The BLM is a decentralized organization, with headquarters and 12 state (often regional) offices. Within the state offices, there are district offices and resource area offices. Most of the daily activities and processing take place at the resource area offices. As a result of the historical land activities, much of the land is in a checkerboard pattern, with BLM land interspersed with state and private lands; this has resulted in the need for close working relationships with state and local jurisdictions, particularly in sharing digital data.

There are specific types of information, for example, the PLSS (the geographic coordinate data base) and the Automated Land and Minerals Record System (ALMRS), that are critical to the BLM that also have other users. Data layers important to the BLM include transportation, meteorology, demographics, geology, soils, and a variety of layers that are specific to given land management agencies depending on the particular local needs.

Defense Mapping Agency

The primary mission of the DMA is to provide mapping, charting, and geodetic (MC&G) products and services in support of worldwide military operations. This responsibility also includes military training activities and the production of military specific products to support weapon and system command and control requirements that cannot be fulfilled by civilian MC&G programs. The DMA activities support worldwide statutory responsibilities such as the production of flight information product (FLIP) charts, aids to navigation and pilotage charts, and related public safety products. A significant percent of their product coverage requirements are met with bilateral agreements through coproduction with other U.S. federal agencies and foreign allies.

The DMA supports the use of the SDTS as a national digital data exchange mechanism, but views the SDTS as an umbrella standard rather than one to be used for exchanging data. Therefore, the DMA is working with the FGDC to establish the DMA's Vector Product Format (VPF) as a federal profile for the SDTS. The DMA is also supporting the Digital Geographic Information Exchange Standard (DIGEST), an international standard that will be used for data exchange among NATO nations. DIGEST accommodates object-oriented data formats (in Appendix A of DIGEST) and relational data formats (VPF in Appendix C of DIGEST). Therefore, a logical step in the standards process now under investigation is to make profiles of the SDTS for both DIGEST formats.

The DMA recently completed a two-phase study that addressed its future specifications and standards (DMA, 1991). The DMA concluded from the results of this study that the trends in future Department of Defense systems were for increased capability for merging (or "fusing") multiple spatial data sets. This trend is the result of supporting users who use diverse computer systems and need different digital products at various scales. The manipulation of spatial data in GIS to meet these needs will be the norm, not the exception.

The DMA completed the installation of its Digital Production System (DPS) in early 1993. This very large spatial data extraction, processing, and production system attempts to minimize the overlap of data collection within the organization. All spatial data needed to satisfy DMA's mission will be stored within a single set of DPS data bases. These data will then be used for producing products at various scales and levels of detail to meet customer needs.

> Federal Profiles to the SDTS
>
> The SDTS is an extremely complicated, detailed format for exchanging spatial data. Because of its complexity and its general applicability to all forms of spatial data, subsets have been proposed to meet specific data exchange requirements of federal agencies. These limited subsets, usually with a much more restricted set of data elements, are identified as federal profiles of the SDTS. Federal profiles are currently proposed for the Bureau of the Census's TIGER files and DMA's VPF and DIGEST.

The major criticism of the DPS is its limited flexibility for fast response to requests for new types of data. Because it was built as a production system, its main focus is to produce standard products that meet defined accuracy, content, and temporal requirements. It does not easily handle requests for ad hoc data and tailored products that are anticipated in the future (this shortcoming is expected to be resolved in the next few years as the DMA ramps-up its production on the DPS).

The DMA maintains spatial data catalogs as hard copy showing the various spatial data holdings, their accuracies and types of data, and their various scales. These catalogs are planned to be put on-line within the next few years.

Although the DMA has expressed interest in supporting the NSDI, their support will be tempered by their primary mission. The needs of the military departments and other federal agencies for spatial data over foreign countries will normally take priority. If the spatial data exchange standards can be accommodated without undue expense, the DMA will likely be a proponent.

Environmental Protection Agency

The EPA has been involved in a very large information integration initiative since 1988. The EPA is currently investing more than $500 million per year in assembling spatial data (usually collected by others) and disseminating the data. Because of this, the EPA has become the largest government consumer of spatial data in support of GIS applications. Although a member of the FGDC, the EPA is not responsible for the production of any spatial data layer of the NGDS, but it does have lead responsibility with the FGDC activity in state and local liaison.

Protecting the environment requires suitable geographic information. It involves the understanding of the spatial relationships of human population centers, natural resources, and sources of pollution. Early in the information initiative, the EPA Office of Information Resources Management recognized the powerful tools that GIS provides to help analyze these three classes of data.

In 1989, the EPA established the Geographic Resources Information and Data System (GRIDS) as the agency's central repository for national spatial data bases. GRIDS provides a query capability that ties together the spatial data bases under the EPA's control. These include the Bureau of the Census TIGER files, the USGS 1:100,000 DLG hydrography and transportation data, the DMA 1:250,000 Digital Elevation Model (DEM) data, the SCS data files, 1:2,500,000 hydrographic basin files, the FWS's National Wetlands Inventory, 1:2,000,000 political boundary files, 1:1,000,000 ecoregions files, the USGS Geographic Names Information System (GNIS), and the Reach files for all U.S. rivers. Requests for information not in these files are sent to the FGDC.

Integration of these data is needed for the EPA to recognize floodplain hazards, wetlands pollution, surface, and ground-water quality problems, potential unstable foundations, population proximity to environmental buffer zones, and many other types of analyses. The integration of the separate data bases has required the development of a number of standards. These standards are expected to overcome the two major EPA barriers to data integration: inconsistent interfaces and no common link between data bases.

A major objective of the EPA approach has been the potential capability of overlaying various data sets. This capability, however, has been slow to materialize. There currently are no standard "tie points" that allow data of different scales and accuracies to be automatically rescaled, thinned or densified, and overlaid on each other with relative consistency. This problem must be solved for the EPA data to be used to their potential.

Even with this problem, the EPA is assembling the most extensive multiagency collection of nationwide spatial data. In doing so, it has developed coordinating mechanisms with FGDC, the Bureau fo the Census, the USGS, the Army Corps of Engineers (COE), the National Geodetic Survey, and the National Institute of Standards and Technology (NIST). The EPA also has a data sharing program with all states and Puerto Rico and a working relationship with the National Center for Geographic Information and Analysis (NCGIA), sponsored by the National Science Foundation (NSF).

The EPA is developing an ambitious program, EMAP, "to assess the nationwide distribution of ecological resources and to assess trends in their conditions. A unique aspect of the program is its reliance on probability-based selection of sampling locations for both of those major goals" (BEST/WSTB, 1992). The MSC found that many aspects of the SCS's NRI, EMAP, and other national statistical surveys appear to overlap and be duplicative; however, because these surveys have each selected different probabilistic sampling methodologies, sharing of data is often inhibited.

The goals of the state/EPA Data Management Program are twofold: to build and maintain the infrastructure for data sharing with the state environmental agencies (basically complete), and to integrate this data across multiple media and programs. Because of the aformentioned problems, satisfaction of this second goal may take some time.

Fish and Wildlife Service

The principal mapping activity of the FWS is involved with wetlands mapping. The FWS is also responsible for coordinating the wetlands data layer of the NGDS. (For additional information pertaining to the FWS see Chapter 6 and Appendix A dealing with wetlands and spatial data.)

The flagship program in spatial data of the FWS is the National Wetlands Inventory (NWI). The NWI was mandated by Congress and established in 1974 to standardize terminology and to inventory the nation's wetland habitats by 1998 (conterminous United States). The NWI has currently produced 20,000 wetlands maps covering 60 percent of wetlands areas of the lower 48 states and 16 percent of the Alaskan wetlands areas. The primary map product is the 1:24,000 scale map that shows the location, shape, and characteristics of wetlands and deepwater habitats. Approximately 3,200 of these maps are produced each year. At this rate the FWS expects to complete the NWI for the contiguous 48 states by 1998. The FWS maps use USGS 7½- or 15-minute topographic maps as their base. Other wetlands maps include 1:100,000 and 1:250,000 small-scale maps, and 1:62,500 or 1:63,360 large-scale maps.

The FWS considers their wetlands data collection to be cost effective and does not want to duplicate data captured by others. However, this is not an easy task. Various federal agencies are involved in wetlands data collection because of their particular mandates and missions.

Wetlands data are supplied to about 50 user organizations by the FWS. Among these, Ducks Unlimited had been paying the FWS to digitize wetlands areas of special interests. The FWS was recently told by the Department of the Interior (DOI) Inspector General (DOI, 1992) to stop this activity and to concentrate on completing the NWI, because the digitization was not included in the original congressional authorization of the NWI. This issue was resolved in 1992 as the NWI authorization was revised (P.L.102-440, §305); the FWS can provide digital products of the NWI providing the requestor pays for 100% of the associated digitizing costs. Because the NWI uses contractors, the flexibility exists to add additional digitizing tasks without affecting the production schedule for the typical NWI products. This limits automation to those that have access to the needed funds for digitizing.

Although the FWS has been digitizing and exchanging digital data for many years, they appear to have taken a more practical approach to exchange standards. Their exchange formats are directly related to the systems they have in house and to the formats of the data they receive from other agencies. Source data sets are acquired from the USGS in the form of 7½-minute quads in DLG-3 optional format, orthophoto quads, high-altitude photography, and Landsat and SPOT imagery. The FWS wetlands inventory data are provided on magnetic tape in multiple formats. (Additional discussion of data and information issues concerning wetlands are in Chapter 6 and Appendix A.)

U.S. Forest Service

The USFS, like the BLM, is basically a land management agency responsible for 191 million acres of public lands. The mission of the USFS (much like the BLM's) emphasizes multiple use and sustained yield. The difference between the roles of the USFS and the National Park Service is that the USFS is multiple land use whereas the National Park Service is usually single use. Also like the BLM, the USFS is geographically dispersed, with 156 national forests and 653 ranger districts. The ranger districts are probably the most important sites that use spatial data. Each district office administers from 150,000 to 500,000 acres and is responsible for virtually everything that occurs in the district.

The USFS obtains its spatial data from in-house activities; other federal, state, and local agencies using cooperative agreements and cost-sharing arrangements; and private companies as a condition of permit or timber sale on USFS lands.

The USFS produces digital spatial data from their base series maps and orthophotography. These base maps use the USGS 1:24,000 base, modified to include land ownership status, administrative boundaries, and transportation networks (at a different definition of the USGS base). About 10,500 quadrangles are required to cover the national forests. The USFS also produces a cartographic feature file in digital form. This uses a simple format and includes coordinates of map corners and point and line features; boundaries are represented as lines (not polygons); the files are not topologically structured but include feature codes; and the nodes are digitized. Although the cartographic feature files are less than optimal for GIS application, it is what the USFS could afford to do; these files can be converted into DLG format.

The USFS has been designated as lead agency for the vegetation subcommittee of the FGDC.

Soil Conservation Service

The SCS within the U.S. Department of Agriculture (USDA) has as its mission to help the states, cities, interest groups, and private citizens to better manage land and water resources. Their charter includes soil surveys, snow surveys, watershed studies, flood control, and conservation technical assistance to landowners.

The SCS long ago realized the benefits of using a GIS for tracking and manipulating spatial data. The SCS uses the public-domain, UNIX-based GRASS system as their primary GIS, and the SCS has tailored interfaces for their personnel.

Currently, the SCS spends about $220 million annually on mapping of soils and other resource information over the United States. This breaks down to $130 million for conservation technical assistance mapping, which includes wetlands mapping (at 1:7,920 scale); $10 million for watershed and subwatershed resource mapping, and $6 million for river basin mapping at 1:100,000 and 1:5,000,000 scales. The remaining $66 million is spent on soil mapping at scales of 1:12,000 to 1:31,680. Soil surveys are a major SCS product, published as printed reports and paper maps but now also becoming available as digital data bases.

The SCS in conjunction with the ASCS and the USGS is proposing a 1:12,000 digital orthophotography program. If implemented, this would considerably increase the store of digital data nationally and at a scale of considerable use by local investigators.

The National Resources Inventory (NRI) is another major data product of the SCS. The NRI contains soil characteristics and interpretations (slope, depth, permeability, salinity, and acidity), land cover, land use, erosion, land treatment, conservation treatment needs, and vegetative conditions. The NRI is used to monitor trends in the status and condition of natural resources and to help formulate state and national policy. Data are collected for the NRI by primary sampling units (PSUs). There are approximately 300,000 PSUs over the United States and data are collected by field units in over 1,000,000 points (three per PSU). The point data are then summarized for statistical purposes and trend analysis.

United States Geologic Survey

The USGS has the responsibility for the National Mapping Program for the United States. This program includes the production of 1:24,000 scale (7½-minutes) quadrangle maps for the conterminous United States, Hawaii, and Puerto Rico and 1:62,500 (15-minute) topographic quadrangle map coverage of Alaska. In addition to this primary map series, the USGS also produces intermediate and small scale map products at 1:50,000, 1:100,000 and 1:250,000 scale in both quadrangle and county formats.

Many of the spatial activities of the USGS have previously been reviewed by the MSC. The report *Spatial Data Needs: The Future of the National Mapping Program* (1990) reviewed user needs for the USGS spatial data products, and the report *Research and Development in the National Mapping Division, USGS: Trends and Prospects* (1991) evaluated the USGS's R&D program. We reaffirm the conclusions and recommendations in these two reports and are encouraged by USGS efforts to be responsive in implementing these recommendations.

The USGS has assigned the National Mapping Division (NMD) with responsibility for conducting the National Mapping Program. In 1990, they completed the first version of the 1:24,000-scale map coverage for the conterminous 48 states and Hawaii, and expect completion of the initial mapping of Alaska by late 1992. Now that the initial hardcopy coverage of the United States is basically complete, the NMD is focusing on ways to maintain the currency, accuracy, and usefulness of their maps.

Many of the federal and state agencies rely on the 1:24,000 quadrangle map series to form the base data upon which other information is referenced. The use of the 1:24,000 quads by other federal and state agencies is one method for reducing duplication of data collection activities. The NMD produces digital spatial data in various formats in response to requests from a wide variety of customer.

In 1987, the NMD undertook a new system development effort called Mark II. This program is to implement advanced technologies and production procedures to satisfy their requirements through the year 2000. The major aspect of this program will be the population of the National Digital Cartographic Data Base (NDCDB) with spatial data representing the content of the 1:24,000 primary map series and other smaller scale series.

In 1991, the NMD changed the focus of the Mark II system development effort to be more of an incremental "commercial off-the-shelf" (COTS) procurement. This re-named Automated Cartographic System (ACS) marries advanced technology developed for other federal cartographic systems with the latest COTS hardware and software. The system is to encompass an open-ended architecture so additional COTS developments may be added in the future. The ACS will be used to support products for the proposed the SCS/ASCS/USGS digital orthophotography program.

The USGS is attempting to accelerate the data collection effort for the creation of the NDCDB. This is being done through data exchange with federal and state agencies, and by contracting out some of the digitizing to private companies. The USGS is working with the USFS for acquiring large amounts of 1:24,000 DLG data. Hypsographic data at 1:100,000 scale in digital form is also being produced by the DMA in a cost-share agreement. Finally, both the USFS and the BLM are developing DEM data to USGS standards under a work-share agreement.

The USGS is also using GIS technology and spatial data analysis for water-resource studies. There is currently a nationwide computer network of over 200 locations staffed with GIS users working on a variety of water-related studies. The system has been used to work with groundwater models, define drainage basins and river systems, assess runoff in urban areas, and examine trends in water quality.

The National Water Quality Assessment (NAWQA) program also has a major influence on USGS spatial data activities. Sixty NAWQA study units will complement the normal 423 long-term water sampling stations to collect data for water-quality studies. Spatial data will be used to classify lands and detect change for tracing pesticides and trace metal concentrations in the environment, and for hydrologic simulation models.

STATE AGENCIES

Spatial data activities typically occur in several agencies in each state. The common agencies include the functions of natural resources, environmental protection, agriculture, transportation, taxation/assessment, emergency management, and planning and community development.

Some state-level spatial data are collected and used at the scale of 1:24,000, and some agencies or programs require larger or smaller scale data. Spatial data are used for environmental applications, though each state's transportation agency has specific requirements related to the road and other transport systems. Large scale maps are also used in the assessment of land values and taxation in a function that is frequently but not exclusively delegated to local governments. There are also various other applications that make use of demographic and economic data.

Current Situation

GIS technology is being implemented in many agencies in virtually every state today (see Warnecke, 1992). The acquisition of spatial data is a very significant percentage of the effort (and resources) for each GIS project. Most state agencies have similar requirements for spatial data and therefore there is significant potential for sharing of acquisition efforts and data. The USGS 1:100,000 and 1:24,000 maps are the most common source for base data; the SCS soil maps (1:15,840) are also frequently used. Conversion of these maps to digital form is therefore a focus for most of the state spatial data development efforts. (Since the USGS also has a program underway to digitize their 1:24,000 maps, there is a strong potential for data sharing with the states in this activity, see Chapter 8.)

Several states have passed or are developing legislation that affects spatial data (see Table 4.5). In some cases the legislation creates or designates a coordinating body. In others it provides access to funding for local governments to modernize and implement land records (see North Carolina example in Chapter 7). Also it creates requirements for spatial data to support a program or regulation. In still others it affects the availability of spatial data developed by state agencies.

Growth management, mining reclamation, redistricting, tax mapping, transportation planning, traffic safety, hazardous materials, natural resources, economic development, and land records modernization are key areas of spatial data activity at the state and local levels.

In the past, most state spatial data activities have been for special programs, often with federal funding and/or direction. However, this situation is beginning to change. States are emerging as an important key player in the collection, dissemination, and coordination of geographic and land information.

One of the major stumbling blocks to the collection of spatial data by state agencies has been funding. To make matters worse, the federal government has delegated more responsibilities to the states (and likewise, states to local government) without matching funds in most cases. In spite of these problems, more and more activity is occurring at state levels to collect and process spatial data using GIS tools.

While much of the state level spatial data acquisition in the past was performed on an individual agency basis, most states now have some level of recognition of the need for sharing and cooperative efforts. Thirty-five or more states have established some type of coordinating mechanism or lead agency for GIS or spatial data. In some states that recognition is formalized by assignment of a lead agency with statewide responsibility or by the formation of a GIS coordinating committee. Some states have implemented legislation and some have prepared or are preparing statewide plans for spatial data development, maintenance, and use. In other states informal arrangements have been established between and among pairs or groups of agencies. In some states, however, cooperation suffers from interagency rivalries, incompatibility between specifications or between systems, or lack of resources to establish coordination mechanisms.

There is a growing awareness of the mutual benefit to be gained by communication among states on this topic. Several meetings and surveys have been conducted in recent years in an attempt to improve the communication among states. At the invitation of Georgia Governor Zell Miller, representatives of 39 states met during the 1991 GIS/LIS meeting in Atlanta. The meeting resulted in the formation of the National States Geographic Information Council to address coordination and common issues within the states. One of the most repeated themes at this meeting is that the federal agencies need to work together more effectively to form stronger spatial data partnerships with the states.

Federal Relationships

Some state programs involve direct cooperation or coordination with related federal agencies and programs. Some others are operated in response to federal requirements for transportation, environmental protection, or other functions.

Where state programs are funded by or are performed in conjunction with a federal program, some level of state-federal cooperation exists. The level of cooperation varies from establishment of standards and sharing arrangements to cooperative efforts for digitizing the USGS quad sheets. Federal reporting requirements affect some state spatial data activities. Both the DOT and the EPA have had an important effect in this area. Many forms of environmental control and regulation emanating from the federal government require the development and use of spatial data, sometimes to federal specifications and others merely to meet federal reporting requirements. In some states specific programs such as the selection of nuclear waste disposal sites (e.g., Pennsylvania) and determination of oil revenues (e.g., Alaska) have been guided by federal programs or regulations. Furthermore, the Bureau of Census' TIGER data forms the legal basis for the most recent redistricting process.

The National Geodetic Survey (NGS) operates a state advisory program in 26 states. The NGS state advisor provides technical assistance and stimulus for improved spatial data activities.

Western states have developed special relationships with federal land management agencies that do have limited activities in the east. In some of these western states there is sharing of data between the state agencies and the BLM and the USFS, though in other states these entities operate independent of each other. In any case, all these relationships appear to be ad hoc and often narrowly defined.

TABLE 4.5 Selected state legislative actions affecting spatial data and information activities.[a]

State	Citation(s)	Type	Description	O	CR	L	F	P	A
Alaska	1900 AK.Ch. 200	Open records	Recent amendments to open records law to take into account nature of automated information		✓	✓	✓	✓	✓
	Alaska Stat. §14.40.095	Education	Establishment of center of information technology at Univ. of Alaska that includes GIS						
Arizona	A.R.S. §37-173 (1989)	GIS organization	Requires Resource Analysis Division to set up GIS	✓					
California	Cal. Gov. Code §6254.9 (1990)	Open records	Excludes computer software developed by government agency from public records and allows sale and licensing of the software						
Colorado	C.R.S.24-72-205 (1992)	Open records	Provides for governmental copyright of public records, and for the charging of fees for information based on the cost of providing the electronic services and a portion of the costs for building and maintaining the system		✓				✓
Delaware	1992 H.J.R. 17	GIS organization	Establishes a GIS oversight committee to coordinate GIS activities and resources	✓					
Florida	1990 Fla. Laws 217	Appropriations	Statewide public lands GIS						
	Fla. Stat. @282.403 (1990)	Data processing	Transfer of data related to growth management among state automated information systems						
Idaho	Idaho Code §39-120 (1990)	Health and safety	GIS for water resources						
Illinois	Ill. Rev. Stat. ch. 111½, par.7056 (1988)	Solid waste management act	Solid Waste Management Act; application of Dept. of Energy and Natural Resources GIS						
Indiana	1990 Ind. P.L. 6	Census	Study costs of acquisition of GIS						

State	Citation(s)	Type	Description	O	CR	L	F	P	A
Iowa	Iowa Code §455E.8 (1989)	Natural resource regulation, ground-water protection	Develop and maintain a natural resource GIS						
Kansas	1990 Kan. SB 793	Appropriations	For state water plan						
Kentucky	KRS §61.970	Open records (public access to gov. data bases)	An open records law directed specifically at GIS data and GIS. Exempts GIS data base from public disclosure if for commercial purpose		✓		✓		✓
	KRS §61.975		Fees for copying GIS, even if not for commercial purpose if nonstandard "product"		✓		✓		
	KRS §61.992		Penalty for abuse of commercial use						
	Senate Concurrent Resolution No. 112	Task force to study certain issues	GIS/GPS task force	✓					
Louisiana	House Concurrent Resolution No. 171	GIS network	Formal recognition of existing task force, and directed to attempt to secure federal funds	✓					
Maine	12 M.R.S. §1753-A, 1756 (1991)	Natural resources	Affects organization and representation of GIS on the Information Services Policy Board, and authorizes the Board to set fees and standards for the Maine GIS	✓	✓	✓			✓
	30-A M.R.S. §4342 (1989)	Planning and land use	Growth management program; development of GIS	✓					
Maryland	1992 H.B. 1538	Automated mapping-GIS	Creates a State Automated-GIS Review Board; authorizes products and services to be made available to the public	✓	✓				✓
Massachusetts	1990 Mass. H.B. 5701	Appropriations	GIS manager						
Minnesota	1990 Minn. Ch. 594	Appropriations	Funds for consultant to study GIS						
	1991 Minn. Ch. 254	Appropriations	To Commission of Natural Resources to develop GIS tools to correlate forest bird populations with dynamics of forest landscape						

State	Citation(s)	Type	Description	O	CR	L	F	P	A
Mississippi	Miss. Code Ann. §19-3-41 (1990)		Provides for state authority to approve local gov. GIS; authorizes borrowing for GIS by local gov. and the imposition of a tax to repay loans	✓			✓		
Nebraska	1991 Neb. L.B. 639	Intergovernmental cooperation	Creates GIS Steering Committee	✓					
Nevada	1991 Nev. AB 772	Congressional districts	Redistricting congressional districts, referencing use of GIS	✓					
New Hampshire	RSC 4-C:3 (1989)	State planning	Statewide GIS development and support for planning purposes	✓					
North Carolina	NC. Gen Stat. §102-17	Official survey base	Financial assistance to counties for base maps				✓		
	1991 N.C. Ch. 285	Public records; special legislation	Qualified exception from Public Records Act for certain GIS. County may require agreement that GIS data not be resold for commercial purposes.					✓	
	NC. Gen Stat. §143-345.6	State departments	Land Records Management Program for deeds, maps,and plats. Creates advisory committee	✓					
	1991 H.B. 356	Public records; special legislation	Creates an exception to the Public Records Act for GIS in Lincoln and Brunswick counties		✓				✓
Rhode Island	R.I. Gen Laws §16-32-30	Education	Univ. of R.I. to establish GIS laboratory, and assist state GIS						
Utah	1991 Ut. SB 21	GIS	Appoints state data processing coordinator; creates Automated Geographic Reference Center which will provide GIS services to state agencies, set policies, manage state GIS data base, set fees	✓				✓	

State	Citation(s)	Type	Description	O	CR	L	F	P	A
Vermont	3 V.S.A. §20 (1989)	Executive branch	To develop strategy for GIS	✓					
	24 V.S.A. 4345	Municipal and regional planning	Develop regional data base compatible with GIS						
	1992 H.B. 955	GIS organization	Authorizes Governor to make Vermont GIS a not-for-profit corporation	✓					✓
Virginia	§2.1-526.18-22, Va. Code		Establishes Division of Mapping, Surveying, and Land Information Systems and the position of Coordinator to coordinate these activities for state and local government	✓					
	1992 H.B. 267	Local government	Provides that local governments may develop GIS and require their departments to use them	✓					
Washington	RCW 43.63A.550 (1990)	Dept. of Community Dev.	Dept. of Community Dev. to study what amounts to GIS to support growth analysis	✓					
West Virginia	1992 H.B. 4433	Public records	Establish requirements for fulfilling requests for records in electronic format						✓
Wisconsin	1989 Wis. Laws 31 and 339	User fees	Empowers local governments to modernize land information; establishes standards within and between governments	✓			✓	✓	✓

O - Organizational structure; **CR** - Cost recovery; **L** - Liability; **F** - Funding mechanism; **P** - Privacy; **A** - Access

[a] Information provided by H. Bishop Dansby, GIS Law and Policy Institute, Harrisonburg, Virginia.

Standards

There has been a great deal of interest and activity in the development of data exchange standards. The current practical reality, however, is that only de facto standards based on available data or formats of software-specific systems are recognized in state and local governments. As a default choice, the USGS DLG (1:100,000) and the Bureau of the Census TIGER formats are often used because data are available at low cost and contain complete coverage of each state.

Problems

Probably the most significant spatial data problem facing states today is a lack of resources and a shortage of funding for data acquisition and management. States often look to the federal government for assistance, particularly with regard to spatial data required for federal mandates. In most cases, however, the federal government is not able to provide the desired assistance.

Coordination of spatial data handling among state agencies has been a chronic problem. Differing requirements, limited resources, and organizational rivalries have impeded cooperation in spatial data activities. This problem is being addressed in many states today and the situation is improving significantly. Even though many states now have a coordinating body and data sharing among agencies is improving, they are still depend on a continuous flow of accurate and timely spatial data.

LOCAL GOVERNMENTS

Local governments are major creators, maintainers, and users of spatial data. Analyses have indicated that as many as 90 percent of local government (approximately 80,000 municipalities, towns, townships, regional planning entities, and school and special districts) operations involve some use of spatial data. Many have stated that they spend more on GIS activities than the federal and state governments combined. Local governments use spatial data for a wide range of activities that include: real estate assessments, land-use planning, public works, water and sewer utilities, resource management, environmental control, solid waste management, emergency management, and health care management.

Local governments generally require medium- or large-scale maps and data. While some local government activities, particularly in rural areas, can be supported by the 1:24,000 scale maps and TIGER data, most local government activities in urban areas require data at scales of 1:4,800 to 1:1,200 (1″ = 100′). Local governments have extensive requirements for cadastral maps since the ownership parcel is a key entity in many local government functions. Planimetric features such as roads, hydrography, and buildings are also important features; soils and forest cover are important resource features. Many local governments operate public works and utilities departments that require data on the location of public facilities and utilities. All of these features must be mapped at medium or large scales depending on the level of urbanization and the parcel sizes.

Current Situation

There has been a major movement to develop and implement GIS among local governments in the past several years. The major effort of each GIS project has been the development of large-scale and medium-scale spatial data bases containing cultural, infrastructure, and resource features. These projects usually involve the creation of new base maps from high-resolution aerial photography and geodetic control. As a result, these projects can take five or more years and cost several million dollars. Because of the resources required there are examples (e.g., Indianapolis) of data sharing among the local government and utility organizations within a geographic area. However, there has been limited cooperative activities between local governments and state or federal agencies.

Some states have established programs of assistance and/or regulations specifying mapping standards for local governments. Most of these focus on equitable tax mapping throughout the state. Notable among these are South Carolina, Missouri, Oklahoma, and New York. In these cases, mapping standards employed by local governments tend to be uniform and of high quality throughout the individual state.

Recently some states, such as Florida, Georgia, and Vermont, have established growth management policies and/or regulations that have placed requirements on local governments for the acquisition and use of spatial data. These requirements are sometimes accompanied by resources for compliance. Wisconsin has implemented the Wisconsin Land Information Program that provides access to state funds when local governments implement a land records modernization plan.

Spatial data development in a local government is now typically incorporated in a GIS program. The GIS program in turn is typically a consortium of several departments across the city or county organization or of multiple city, county, and utility organizations. Because this can cross a multitude of jurisdictional boundaries, these local projects require considerable organizational cooperation. The lead agency is usually a GIS user department such as planning or public works. GIS data activities have traditionally taken place outside of the conventional information services department.

Federal Relationships

There is little sharing of data between local governments and federal agencies. In recent years there has been very little federal funding provided to local governments and so little sharing has taken place.

The FEMA flood insurance program has generated maps of flood zones that are used extensively by local governments. These maps are required for definition of locations for flood insurance purposes. Even though the scales and formats of these maps are often incompatible with other local government spatial data, they are used out of regulatory necessity.

Federal regulations significantly affect local governments. EPA regulations in particular are generating requirements for digital data among local governments. For example, the recent EPA regulations (e.g., the non-point discharge elimination standards) are stimulating many local governments to collect spatially referenced data for purposes of storm water pollution permitting. This requires the local governments to integrate geographic, water, and topographic features with other local spatial data.

Standards

Generally the same standard issues facing states (above), face local governments. In fact the issues may be exacerbated by the often dramatically different scales used at the local level.

Problems

Local governments suffer from the same lack of funding and resource problems as states. The magnitude of the effort required to acquire spatial data for a local government is often a serious problem. While new technological tools and procedures are being developed, the development of a spatial data base continues to be a major concern.

There is about the same level of communication or coordination of spatial data activities among local governments as there is with state and federal agencies. Within a specific area, such as a county, there may be a cooperative GIS project that involves multiple organizations, but there are very few effective mechanisms for communication with other local governments to obtain guidance or resource sharing. A few states (e.g., Wisconsin) have a program for guidance or technical assistance from the state to local governments.

PRIVATE SECTOR

The private sector makes broad use of spatial data technology and has done so for over two decades; they are both producers and consumers of digital spatial data. Many uses involve digitizing proprietary single-purpose maps—for example, an electric utility company computerizing its "outside plant" facilities: utility poles, rights-of-way, and transformer substations. Timber companies were early adopters of GIS technology to inventory and analyze timber stands. While these applications involve spatial data, in many cases they result in the acquisition of proprietary and/or special purpose spatial data that are not of potentially broad usefulness within the national spatial data infrastructure. One exception is the sale of some of the spatial data by telephone companies to local governments as part of the enhanced 911 emergency service.

On the other hand, many private sector applications call on the same generic spatial data sets time and time again, specifically: boundaries or centroids of statistical areas (counties, census tracts, or zipcodes) and street centerline spatial data sets.

Typically, a person (or small group) at a private company becomes aware of a GIS or desktop mapping package, sees that spatial data technology could be useful to the company, and buys a copy of the product. They rapidly discover that the new system cannot deliver any value without a copy of an appropriate spatial data set. Practically without exception such a person assumes that the spatial data set is available from the government at low cost. What they discover is that either:

- the appropriate spatial data set (census tract boundaries, for example) is available only from the private sector;
- a government data set like the Bureau of the Census' TIGER file is sufficient for many of the applications and it can be acquired from a private company cheaper and/or in a more convenient ready-to-use format;
- a proprietary enhanced version of a government spatial data set is necessary to make the application work, because of some shortcoming of the government product; or
- the required spatial data set does not exist and the company must have it built from scratch at a substantial cost.

Incidentally, this experience is not restricted to private companies; in many cases government agencies at all levels can only find the spatial data sets they need from private suppliers.

The private sector has many advantages over government in use of spatial data technology for several of the following reasons:

- Private applications are generally simpler or at least more focused: the impact of spatial data technology on their bottom line.
- Private sector procurement practices are simple and streamlined, so companies have better access to the latest technologies at lowest cost.
- Private sector decision making is streamlined; action can be taken quickly.

In addition, private sector spatial data vendors in the United States have the tremendous advantage—practically unique in the world—of unrestricted royalty-free usage of federal spatial data sets like TIGER and DLGs, as well as copyright-free access to federal paper maps as a basis for proprietary spatial data sets.

As a result of these advantages, a robust private sector community of users and vendors of spatial data technology has evolved, which consists of a large group of users, software vendors and data vendors. Users represent nearly every sector of business: fast food chains selecting new sites, catalog marketers targeting customers, taxi companies dispatching cabs, insurance companies evaluating hurricane risks, food wholesalers routing delivery trucks, sales managers delineating sales territories, phone companies planning communication infrastructure for future developments, and banks complying with home mortgage disclosure regulations. Every one of these application sectors is a consumer of the spatial data infrastructure and is expected to grow for years to come.

There is a growing number of software vendors offering a range of products from GIS, desktop mapping, desktop marketing, routing, and dispatching. In every case these software systems are useless without access to a spatial data set. Increasingly, the customers of software vendors demand "plug and play" spatial data sets; they want to concentrate on getting their jobs done without the costly diversion of digitizing or massaging a spatial data set. Many vendors only break even on sales of their software packages but make profits reselling spatial data sets to their customers.

Spatial data set vendors are either companies that simply transform data sets like TIGER or DLG to the internal representation required by systems like AutoCAD, or companies with more ambitious goals of creating proprietary spatial data sets from scratch or by significant investment to improve a public data set like TIGER. The latter companies have specialized to serve geodemographic markets or emerging IVHS (intelligent vehicle-highway system) applications like in-vehicle navigation and route guidance.

It is significant to observe that private spatial data set vendors dominate the supply of spatial statistical area boundary data bases as well as street centerline data sets. In aggregate, the combined data base creation and maintenance budgets of these companies could exceed federal spending on these data sets. In a previous report of the MSC (MSC, 1990), it was stated that de facto control of significant spatial data sets could default to the private sector. This has probably come true to some degree considering commercial activities of land title companies and the extent that private companies have leveraged off the running start provided by the Bureau of the Census release of TIGER in 1989.

Traditional private sector advantages of efficiency and competitiveness have served the government well when, for example, building of portions of TIGER was put out to bid in the mid-1980s. The private sector as a whole responded aggressively and positively to spatial data set technology with dynamic growth in all three sectors defined above. But private information businesses tend inexorably to monopoly or at best oligopoly. At present, a competitive market exists for some spatial data sets, and the rapid growth of "Business GIS" indicates that value is being delivered at current prices.

As we say several times in this report, what is missing from the federal perspective is a vision of spatial data use beyond fulfilling missions of individual agencies. The value of spatial data technology in promoting productivity or stimulating commerce is not a major concern of the FGDC or of most of its constituent agencies. Expansion of the vision of spatial data use beyond federal agencies' needs will be considered as a future subject for a Mapping Science Committee study.

ACADEMIA

Academia serves primarily in a support role for the improvement of the National Spatial Data Infrastructure. Although the nation's colleges and universities have extensive personpower, GIS experience, and specialized spatial data sets, there are few programs that coordinate their efforts for input to the NSDI. In almost all cases, academia must respond to the requirements of those agencies supporting their research. It is clearly not the role of academia to structure the NSDI, although the productivity of many of its scientists would be enhanced if the NSDI were more robust.

Many specialized GIS research and teaching laboratories have recently been established in U.S. universities. In many cases these laboratories work closely with state-level institutions to assist their states to rationalize and improve their spatial data and to educate students in the methods and techniques of processing and using spatial data. Examples of leading university activities include the National Center for Geographic Information and Analysis (NCGIA—a consortium of the University of California at Santa Barbara, the State University of New York at Buffalo, and the University of Maine at Orono), the Center for Mapping at the Ohio State University, the University of South Carolina, and the University of Wisconsin—Madison. Research programs at these and other universities are advancing basic knowledge in the field of geographic information and analysis. Examples include analysis of error in spatial data bases; the use and value of geographic information; development of new data gathering techniques such as Ohio State's GPS van (Bossler *et al.*, 1991); remote sensing research and others. The NCGIA receives its principal support from the NSF along with additional support from the USGS, the EPA, and others; the Ohio State University's Center for mapping receives funding from the National Aeronautics and Space Administration (NASA), the USGS, and other state and federal agencies; and the University of Wisconsin—Madison is supported by the U.S. DOT, the NSF, the USDA, and various state, local, and private organizations.

Additional research is critical. The recommendations in the MSC's (1991) previous report on *Research and Development in the National Mapping Division, USGS: Trends and Prospects* outlines work that is applicable not only to the USGS but also every research organization that is part of the NSDI. We reiterate our emphasis for continued research on standards.

The explosive growth of spatial data handling technology and applications has whetted the appetite of users for new and improved capabilities to analyze, model, and apply the data to meet their needs. The private sector seems to have satisfied some of these needs through the development of new hardware platforms and software to process the data. Still other needs or desires require innovative R&D to extend data handling and modeling capabilities. Although some results are several years away from the marketplace, the private sector has demonstrated its rapidity to incorporate new spatial data handling advances into commercial products. To realize fully the potential benefits of new technology, researchers in government, the private sector, and the academic community should work synergistically.

REFERENCES

BEST/WSTB (1992). *Review of EPA's Environmental Monitoring and Assessment Program (EMAP): Interim Report*, Board on Environmental Studies and Toxicology (BEST) and Water Science and Technology Board (WSTB), National Research Council, Washington, D.C., 25 pp.

Bossler, J. D., C. C. Goad, P. C. Johnson, and K. Novak (1991). GPS and GIS Map the Nation's Highways, *Geo Info Systems 1*(3), 26-37.

DMA (1991). *Digital Products Study: Uses, Standards and Specifications*, Defense Mapping Agency, Fairfax, Virginia.

DOI (1992). *Audit Report: National Wetlands Inventory Mapping Activities, U.S. Fish and Wildlife Service*, Office of the Inspector General, Department of the Interior, Report No. 92-I-790, 54 pp.

FGDC (1991). *A National Geographic Information Resource: The Spatial Foundation of the Information-Based Society,* Federal Geographic Data Committee, First Annual Report to the Director of OMB, 10 pp. plus 41 pp. of appendixes.

FGDC (1993). *Manual of Federal Geographic Data Products*, Federal Geographic Data Committee, Washington, D.C.,

MSC (1990). *Spatial Data Needs: The Future of the National Mapping Program*, Mapping Science Committee, National Research Council, National Academy Press, Washington, D.C., 78 pp.

MSC (1991). *Research and Development in the National Mapping Division, USGS: Trends and Prospects*, Mapping Science Committee, National Research Council, National Academy Press, Washington, D.C., 63 pp.

OMB (1988). OMB Bulletin 88-11, 1988

Warnecke, L. *et al.* (1992). *State Geographic Information Activities Compendium*, Council of State Governments.

5
SPATIAL DATA AND THE URBAN FABRIC

INTRODUCTION

The basic premise of most federal mapping programs has not evolved as the nation has changed from an economy based on natural resources to one dominated by personal services, manufacturing, and industry. The nation grew from its knowledge and exploitation of natural resources. Crossing the continent required (and resulted in) a knowledge of the physical and biological geography. The mapping of its topography gave government and private industry a context for great plans and small initiatives; the mapping of its geology, a sense of wealth and the location of building materials; the mapping of its soils, an assessment of the potential to feed itself and others; and the charting of its waters, the ability to navigate, generate power, and harvest fish.

As the nation has matured, these mapping needs have lost their priority though not their utility. Delivery of human services, assessment of environmental conditions, responses to natural hazards such as earthquakes and hurricanes, deployment of law-enforcement personnel, education of children, economic development, transportation of goods, delivery of services, and the electoral process itself demand a broader and more detailed information set: one that defines, characterizes, and embraces the urban fabric of a once-rural nation.

The federal focus on mapping natural resources, almost to the exclusion of all else, fails to recognize the growing needs of the federal agencies for information more commonly referenced to street addresses, voting precincts, and land parcels. We refer to these needs within the context of an urban fabric, a fabric that includes both urban and rural areas and consists of information concerning the built and human environment versus one dominated by natural resources. The expenditures for acquiring and encoding information defining and pertaining to places are made in an ad hoc manner and redundantly by federal agencies, local governments, and the private sector. Thus, what could be a national *investment* frequently goes no further than an expense.

We divided this fabric into the following four pieces:

- the land base and use of land,
- the ownership of the land,
- the transportation (street) network that serves it, and
- the addressing schemes that provide common geographic reference.

We discuss these pieces in the above context.

BACKGROUND

The point of departure for this discussion is the findings and recommendations in the report—*Spatial Data Needs: The Future of the National Mapping Program* (MSC, 1990)—that digital, spatial data bases be developed to serve both public and private needs of the country. We studied the case of the urban fabric to consider the needs of urban areas for digital spatial data, to determine how well currently available systems for supplying and maintaining spatial data meet these needs, and to determine how the availability and accessibility of urban spatial data in the future might be improved.

An urban spatial data base is a digital system that defines location and/or spatial references (addresses) of objects, people, and events. The committee identified four general kinds of urban spatial data. This chapter addresses the potential contribution of each toward meeting the needs for an adequate spatial data infrastructure for urban areas of the United States and the current status of each one. In Chapter 9 we present recommendations for technical, organizational, and institutional changes and for cooperative activities to meet the needs for urban spatial data in the next decade. The four components of an urban spatial data base are (1) the urban land base system (planimetric and topographic); (2) the cadastral system; (3) the street centerline system; and (4) the political and administrative boundary system (not further discussed).

In some urban areas, government and private organizations are cooperatively developing GIS. However, individual organizations usually develop their own GIS to respond to their needs. No commonly accepted standards for geographical accuracy, content, format, or transferability of data exist. Consequently, sharing information between different organizations as well as relating information from different organizations for common geographical areas is difficult and sometimes impossible.

The federal government also needs spatial information within urban areas for many of its tasks. Its access to this information is difficult, if not impossible, when so many different systems for collecting, updating, and maintaining geographical information exist and when the terms of accessing that information differ from one urban area to another and from one organization to another.

LAND BASE SYSTEMS COMPONENT

Description

Land base refers to activities and data sets (both analog and digital) associated with defining the location and extent of physical (natural and manmade) features on the Earth's surface. The land base typically incorporates:

- a common coordinate reference system,
- the portrayal of physical features in a planimetric and orthogonal view, and
- the representation of topographic relief and features.

Land base systems include:

- techniques for collecting information, through aerial photography and remote sensing, about the location, extent, and nature of physical features;
- techniques for representing this information graphically and textually;
- linkages to other descriptive information about the features;
- techniques for replacing and disseminating information, e.g., maps; and
- a process for maintaining the temporal currency of the features.

The scale, accuracy, and feature representation, as well as the level of textual and statistical detail characterizing each feature, are varied with little consistency and no standards between the activities and products of similar organizations. However, land base maps are typically the foundation of most spatial information systems, serving to register other layers or themes of information to a particular geography.

Primary Responsibility

No organization or level of government is responsible for building, maintaining, or setting the standards for land base systems. Because of their central import to analog and digital spatial information systems, many organizations have traditionally developed base maps to support their traditional requirements; other organizations have informally adopted (and sometimes modified) the base maps of another organization to satisfy, if only partially, their requirements.

The map products of the USGS and the SCS have served, most frequently, as the base map for federal mapping activities. The traditional natural resource orientation and the scale of the products limit their utility and fail to serve agencies with an urban applications orientation. The USGS and SCS map products are also used extensively by state and local governments and the private sector, with comparable limits on their utility.

The land base mapping of state governments is generally more diffused than that of the federal levels. One consistent exception is the mapping activities of the states' departments of transportation. State agencies rely heavily on federal sources for large-area mapping and to a significant yet lesser extent on the mapping activities of local governments.

Land base maps with wide ranges in type, accuracy, currency, and content are a common product of local government. The land base is constructed by using many different techniques and standards, frequently relying on the private sector as a supplier of some or all of the sources referenced above.

Land-base mapping by the private sector occurs both in support of federal, state, and local government activities and as a component of private business, most frequently in land acquisition, site development, resource management and development, and providing and servicing utilities and transportation networks.

Needs

Tens of thousands of organizations are building and maintaining land bases in the United States. The cooperation and coordination in building and maintaining the land base systems rarely go beyond ad hoc relationships within organizations and between organizations interested in the same geography. The ramifications are obvious: lack of consistency between similar or identical data sets, redundancy, lack (gaps) in coverage, and lack of temporal currency among others. The cost to the national economy in terms of productivity and misdirected revenues is extraordinary.

There are no systematic or programmatic opportunities other than isolated and local arrangements to facilitate the sharing of responsibilities in developing land bases between the various levels of participants: federal, state, and local governments and the private sector. If indeed there is a national benefit to be derived from recognizing and building a better NSDI, surely the institutional processes and programs for sharing responsibility as well as data must be established.

Models

We are not without models—limited in coverage though they may be—that might serve as points of initiation. Among them are:

- cooperative geologic mapping programs of the USGS and state geological organizations (see P.L.102-285);
- the high altitude aerial photography program of the USDA, the USGS, and state governments;
- the cadastral mapping program and the previous funding incentives of North Carolina and county tax assessors;
- the funding incentives for land record modernization in Wisconsin; and
- GIS consortia such as in Louisville, Kentucky; Knoxville and Nashville, Tennessee; and Indianapolis, Indiana.

Rationale for Federal Involvement

Federal needs justify an initiative to reach a more cooperative and less redundant spatial data collection process. The benefits derived by other levels add to the public good and national productivity.

The needs of the federal government span all levels of the geography and detail of land base systems. At the same time, the federal agencies are not always best positioned to collect, validate, maintain, or aggregate information required by federal agencies. The examples are many, including locations of buildings, construction of secondary roads, movement/migration of people, and siting and characterization of public and privately owned utility facilities.

THE NATION'S CADASTRE

Introduction

As William Chatterton (1991, past chairman American Bar Association Committee on Land Records Improvement, personal communication) noted, markets are now linked globally through sophisticated communications technology and have blended into one global trading system. Globalized production and marketing require capital to flow easily across national boundaries. A recent Nobel Prize for economics went to the pioneers of the efficient-market theory in which prices quickly reflect information about property. The work of these Nobel Prize winners shifted the focus to getting information to market as quickly as possible. The growing volatility of the world capital market is a sign that the old system is increasingly going out of control.

The savings and loans crisis and the well-publicized problems in the banking and insurance industries similarly tell us that old safety mechanisms, designed to maintain financial stability in a world of self-contained national economies, are as obsolete as the world they were designed to protect. Even minor glitches in telecommunications and computer systems now have the potential to wreak major havoc in the financial marketplace worldwide.

Transactions in land are the touchstone of a market economy. Public records should, therefore, enable a prospective landowner, whether he is located in New York, Los Angeles, Tokyo, or London, to determine, quickly and unambiguously, the rights, responsibilities, and risk associated with owning a specific tract of land. We already see that land information systems are being driven by the requirements of the market. Most land transactions in the United States now require title insurance, so that the multibillion dollar market in mortgages can at least be assured, to the extent of the mortgage title policy, that it has a first lien on the premises.

The United States lacks what is commonplace in most of the developed nations and many others: an integrated and comprehensive national land records system. Historically, the responsibility for maintaining land ownership information (i.e., a cadastral system) has been delegated to local governments, with state intervention and coordination occurring only limitedly and inconsistently.

Description

Cadastral system refers to those activities and data sets associated with parcel-based land information (McLaughlin and Nichols, 1987). It incorporates:

- a common definition of the parcel and a unique primary parcel identifier,
- a cadastral mapping program with geodetic control, and
- linkages to a series of records.

Information contained in the cadastral system include:

- data about how the parcel itself was created and any subsequent changes,
- core cadastral data from primary sources (e.g., ownership data from the registry office, value data from the assessment office, etc.), and
- other data that may be referenced to the parcel (such as building and improvements data).

Cadastral information may be related to other spatial data in a variety of ways, including

- assigning coordinates to the cadastral parcel,
- registering the cadastral map to a base mapping system, and
- relating the parcel identifier to other indexes (e.g., street addresses).

Primary Responsibility

The states have the primary responsibility for building and maintaining cadastral systems. They have the constitutional mandate for administering the system of real property law and for matters pertaining to the planning, development use, and taxation of land. They are also responsible for establishing the administrative arrangements (such as registry offices) required to support this mandate.

Much of the daily responsibility has been traditionally delegated by the states to local government. Currently, there are approximately 100 million parcels of taxable real property. The records for these parcels are maintained by approximately 80,000 state and local government entities (including counties, municipalities, towns, townships, etc.).

A few states, such as North Carolina and Wisconsin, have undertaken to develop management programs for county land records resulting in statewide data bases. To date these programs have been primarily concerned with developing uniform standards and with assisting (through funding and technical assistance) in local land records improvement programs. North Carolina, for example, instituted a land records management program in 1977, and developed cooperative programs with more than 80 of its 100 counties (see Chapter 7).

Rationale for Federal Involvement

The federal government is responsible for a diverse group of mandates and functions that require parcel-based information. These include aboriginal land tenure; the federal government's significance as a land owner; its role in real estate and asset/facilities management; its role in acquiring property for specific projects; various taxation roles; its regulatory role with respect to real estate financing, interstate commerce, agricultural support programs, environmental assessment, hazardous waste management, etc.; and civil defense and emergency preparedness roles.

The Problems

The problems associated with the existing cadastral arrangements have been extensively documented in studies over the past 30 years. The basic record-keeping systems are often archaic, based on legal principles predating the U.S. Constitution (the rudimentary deed registry concept). Information may or may not be available for a specific parcel (e.g., registration of deed is generally not compulsory) and may be distributed among a number of public agencies. There is generally no systematic procedure for efficiently retrieving the data, and current cadastral index maps are generally not available. The quality of the data stored and their legal validity are often unknown. The ability to integrate parcel data from different sources is generally very limited. The costs associated with these problems show up in the transaction and regulatory costs of the real estate marketplace, the assessment and collection costs of the taxation system, the cost of the record-keeping itself, and others.

Responding to the Problems

Efforts to respond to these problems date back more than 30 years and include proposals aimed at improved survey standards and coordination of parcels, administrative and legal improvements to the deed registry system, automation of land records, development of cadastral mapping programs, integration of parcel data bases, etc. In the early 1980s, many of these proposals were bundled together under the label of the multipurpose cadastre concept, as described in two reports prepared by the National Research Council (Committee on Geodesy, 1980, 1983).

The impact to date has been modest. Although many local governments have invested in specific land record improvements (especially the introduction of computers), a much smaller number have undertaken more comprehensive reforms. The reasons for this limited impact are many and complex but are much more a function of policy and institutional concerns than any underlying technical or financial issues.

The most recent effort to deal with the cadastral problem from a national perspective has been the Department of the Interior study (DOI, 1990) on land information mandated by the Federal Land Exchange and Facilitation Act of 1988 (P.L.100-409). Essentially, it once again reviewed the problems associated with land information management in the United States (with an emphasis on cadastral records); examined a number of recent initiatives at the federal, state, and local levels; and called for a nationwide land information management system. The report's recommendations have gone unheeded. The timing of undertaking land records reform will never be better because the technology is ripe and the task becomes more complex with time.

STREET CENTERLINE SPATIAL DATA BASE

Description

The transportation network is the third and increasingly important component of the urban information fabric. Its importance transcends the utility of a common street map because it is a base for defining, organizing, and accessing places (and their associated information) within both complex urban environments and rural areas.

In a digital format the importance is magnified and its uses are expanded manyfold. What might have served as a guide or descriptor of pathways through a particular geography can now serve as an index to large volumes of tabular and statistical data, as a framework for depicting those statistics spatially, and as an analytical tool.

In its basic form, a Street Centerline Spatial Data base (SCSD) is a computerized street map, where streets are represented as centerlines and the characteristics of the streets (the attributes) are appended. The practice that has developed from almost three decades of experience has led to records that uniquely identify each segment and intersection of the street network; differentiate between the left and right side of each segment and anomalies such as cul de sacs; encode street names, zip codes, census geographies, and other area descriptors; provide address ranges for both sides of the street segment; and incorporate coordinate references and scale at various levels of accuracy, among other characteristics.

Significance and Applications

A national SCSD with these characteristics, if kept current and accurate, would support a wide variety of administrative functions by all levels of government and would also provide a basic spatial framework for extensions to serve a wide a variety of other applications of commercial importance. Applications include facility site selection e.g., schools, health facilities, and transit nodes), socioeconomic planning studies using census data, legislative redistricting, and analysis of demand for and supply of human services based on administrative records address-matched to areas with census data. Operational Intelligent Vehicle Highway System (IVHS) applications of SCSD in logistical operations (e.g., fleet management) promise to reduce driving time and fuel usage. Direct benefits extend to increased productivity of professional drivers and commuters alike, reduced emissions from burning fossil fuels, and reduced dependence on foreign energy sources.

Current Status of Street Centerline Spatial Data Bases

Although no single national data base fills all requirements for an SCSD, products from three agencies provide part of the content of a national street-centerline resource.

USGS DLG Program

During the 1980s the USGS embarked on a program to produce DLGs corresponding to most nationwide paper map quadrangle publications. The 1:100,000 scale DLG series was completed in the 1980s and served as the coordinate and linework basis for nonmetropolitan portions of the Bureau of the Census TIGER data base (see below).

The transportation layer of the 1:24,000 DLG series in April 1992 existed for 5,360 of the 55,000 quadrangles. This series is important because its coordinate accuracy is sufficient for many SCSD applications, and it is derived from the largest-scale (most detailed and accurate) uniform map series available in the United States. Despite coordinate accuracy, the data in the series are limited in applicability because they lack street names and address ranges and are infrequently updated, the coverage is limited, and the accuracy levels are inadequate for many uses.

U.S. Postal Service's ZIP+4 Data Base

Although technically not an SCSD, the U.S. Postal Service's (USPS) ZIP+4 file is an important national resource. ZIP+4, the most detailed postal data base, relates all U.S. addresses to nine-digit numeric codes, each of which usually ties a small number of addresses to one side of a city block. The first five digits are the familiar zip code, which corresponds to a postal delivery unit. Each zip is divided into about 40 sectors (the next two digits), each of which can have 99 segments (the last two digits). In contrast to DLGs and TIGER, ZIP+4 is updated monthly to permit large-volume mailers to precode mailing addresses with postal codes. Mailers benefit from postal rate discounts; the USPS gains by streamlined bulk-mail processing.

Unlike DLG and TIGER, ZIP+4 is not organized as an SCSD. Each logical record in ZIP+4 can define a range of addresses, part of a rural carrier route, a box number, or even an individual person or company. The ZIP+4 file contains no spatial coordinates or census codes. Yet, the ZIP+4 is an important resource to the national SCSD because (1) it is a complete definition of mailing addresses throughout the United States; (2) addresses usually are the most prominent geographic identifiers in people, property, or event-oriented data files; (3) it is updated monthly, making it by far the most current geographic reference data base; and (4) it is widely used as the standard for street naming.

TIGER

TIGER (Marx, 1990), a by-product of the 1990 census, is the closest of the three examples to a broadly useful SCSD. It covers the entire country and is the authoritative repository of the 1990 census geographic definitions. Although TIGER's coordinate content is immense, accuracy and graphical appearance vary because of past cartographic and data processing practices. TIGER coordinates are useable for many applications but fall short of the consistent accuracies of the 1:24,000 DLGs and the needs of many users. Another deficiency of the TIGER files is that only streets in urban areas carry address range and ZIP code attributes. Consequently, only 55 to 60 percent of U.S. addresses are represented in TIGER, which severely limits TIGER's utility. Moreover, as TIGER is a work product of the 1990 census, a schedule of periodic updates is not published.

Problems Caused by Deficiencies of Available SCSDs

A principal problem with the present situation is that there are three spatial data bases, developed to serve the needs of their respective agencies, each containing data that are not easily coordinated with data from the others. Levels of spatial accuracy differ, one (ZIP+4) contains no spatial coordinates, and the temporal updating of information is different and uncoordinated for the three sources. Because many user needs are not met by any of the above sources, a second problem is created when users of SCSDs frequently add attributes of interest by themselves to one of the existing SCSDs without consulting with or cooperating with other public or private users. The result is a costly duplication of effort and lost opportunities for acquiring a superior data base.

Although designed to facilitate specific agency internal operations, both TIGER and ZIP+4 are widely used in a broad range of state, local, and commercial applications. Again, inconsistencies persist. The Bureau of the Census actively markets TIGER and encourages TIGER use with workshops. TIGER is used for marketing studies, dispatching taxis, making inventories of municipal street signs and streets, and routing fleets of vehicles, for example. On the other hand, the USPS uses copyright restrictions to discourage use of ZIP+4 for purposes other than preparation of bulk mailings. Nevertheless, information from the ZIP+4 data base is being tested by at least one state as a mechanism for allocating state income tax to school districts and by marketers for demographic analysis of customers and prospects.

Because versions of an SCSD are being developed independently, little thought is given to issues of standards and formats. The flow of information is now dependent on mandates and missions of specific agencies, legal and market considerations, budgets and profits, poor communication linkages (organizational, personal, and technical), and generally the lack of coordination and cooperation. Adding to the complexity and inconsistency, several private companies have created proprietary data bases combining the best data from TIGER and ZIP+4 and proprietary data sources.

Need for Greater Coordination or Consolidation

There are inter- and intragovernmental opportunities for shared development, collection, and use of spatial data. Additionally, there are valid roles and potential for private sector participation in cooperative efforts as well as in providing value-added products. As an example, information can gain value through the accumulation of incremental improvements, features, detail, and validation or definition of accuracy as it is passed from one governmental level to another or from the governmental bodies to the private sector and on to the end user. The added value is as much a function of the entity having the resources and the mandate (whether legal or profit motivated) to contribute.

With foresight, the development life cycle of information can be used to add value and accessibility to information. Moreover, the design of the information resource can become dependent on the interactions of the various participating levels and with coordination the actions of any particular level can be maximized, realizing the potential efficiencies of a robust NSDI.

There is a need to find the resources to forge the permanent cooperative arrangements that will be required to produce and keep current the SCSD. Many application areas have special needs that are not covered by the current SCSD. For example, TIGER is still missing the kind of attribute information required for transportation network modeling and logistics. The most important are direction of flow, turn restrictions at intersections, travel times on segments by time of day, and turning penalties at intersections. In addition, there should be a minimum set of transportation-related attributes. State departments of transportation could add their route identifier and beginning and ending milepoints to each segment in the SCSD so that they can more easily reference other attribute information, such as pavement condition, accidents, and signs.

Conclusions: The Need to Ensure Access, Use, and Maintenance

An urban SCSD is an essential component of the urban information fabric. There is a clear federal need for a city block/street-level spatial data infrastructure, evidenced in support of the missions of the Bureau of the Census and the USPS. Yet each agency continues to maintain its data base independently.

Clearly, no single federal agency has responsibility for the development and maintenance of a fully functional SCSD. Several agencies have responsibility for pieces of information contained in the SCSD but, like a jigsaw puzzle that is never put together, none has the responsibility nor the duty of ensuring that these pieces fit together. The only two nationwide data bases (ZIP+4 and TIGER) are byproducts of the operational missions of their organizations. Furthermore, the ZIP+4 data base is copyrighted by the USPS to control distribution of outdated copies, the use of which could snarl mail processing. The only general-purpose data base—the 1:24,000 DLG—is incomplete and lacks key data fields required for broad usage. Furthermore, none of these agencies has plans to augment data bases with information about traffic-flow restrictions, which would support future IVHS applications and other logistical operations such as dispatch routing, which has substantial commercial value and utility to governmental units of all levels.

There is a natural but undefined opportunity for sharing spatial data through the entire domain of federal, state, and local governments and the private sector regardless of the specific geography involved. The commercial utility of an SCSD is great enough to encourage several private companies to fund independent multimillion dollar projects to improve public SCSD resources, turning them into proprietary data holdings. Moreover, each of these developers, collectors, and users of spatial information is making considerable investments that mask significant opportunity costs because of redundancy of effort, undeveloped opportunities for efficiencies, and insufficient financial resources available to any one sector.

REFERENCES

Committee on Geodesy (1980). *Need for a Multipurpose Cadastre*, National Research Council, National Academy Press, Washington, D.C., 112 pp.

Committee on Geodesy (1983). *Procedures and Standards for a Multipurpose Cadastre*, National Research Council, National Academy Press, Washington, D.C., 173 pp.

DOI (1990). *A Study of Land Information*, Department of the Interior, Washington, D.C., 61 pp. plus appendixes.

Marx, R. W., ed. (1990). The Census Bureau's TIGER System, Special Issue, *Cartography and Geographic Information Systems 17*(1), 133 pp.

McLaughlin, J., and S. Nichols (1987). Parcel-based land information systems, *Surveying and Mapping 47*(1), 11-29.

MSC (1990). *Spatial Data Needs: The Future of the National Mapping Program*, Mapping Science Committee, National Research Council, National Academy Press, Washington, D.C., 78 pp.

6
SPATIAL DATA AND WETLANDS

INTRODUCTION

One of the objectives of this report is to identify what could be done better or more efficiently if the content, accuracy, organization, and control of spatial data were different. To answer this question it is important to understand the complexity of such an infrastructure, including the spatial data themselves. Spatial data consist of digital representations of geographic objects or features on the surface of the earth including tangible man-made objects such as roads and administrative features such as land ownership parcels that are part of the rural and urban fabric discussed in the previous chapter. An equally important part of a spatial data base, however, consists of features that occur as the result of physical, chemical, or biological processes. To focus on those naturally occurring features, the MSC studied existing programs that generate spatial data bases of wetland features. Wetlands were chosen because they represent environmental aspects of the Earth that need to be delineated, measured, represented, analyzed, and displayed quite differently from man-made objects, such as streets. Wetlands were also chosen because they are of considerable national interest and debate. For example, it can be argued that a federal mandate for "no net loss" of wetlands relies on a consistent spatial data base. Given the fact that wetland data are collected in numerous federal, state, and local agencies as well as several private institutions, the MSC believed that a case study with a focus on wetlands would provide an excellent example of the needs and challenges facing the development and implementation of a robust NSDI.

Maps are graphical models or simplified views that represent the distribution of a collection of features on the Earth's surface. To provide a consistent spatial data representation of a feature there must be agreement on how to classify it, how to delineate it, the criteria for inclusion, valid attributes to assign to it, and the appropriate way to symbolize it. In the previous chapter we examined the federal response to the creation and maintenance of a data base of street centerlines for the nation. Such a file is composed of a relatively unambiguous and easily identifiable set of geographic features. These features are very tangible and can be graphically represented as a set of line segments that form the links of an integrated network. For example, while differences of opinion exist among agencies regarding the appropriate classification of roads and the set of attributes that should be assigned to them, the task of delineating and representing them is straightforward and can be achieved from a wide variety of source materials. In contrast, the natural environment presents a much more complex set of challenges. Naturally occurring features such as wetlands are not only difficult to classify, but they also have poorly defined edges, tend to change through time, appear and disappear at different scales, and often can only be delineated with field work and skilled interpretation based on specialized source materials, such as color infrared photography. Furthermore, the areal extent or quantity of natural features has important ramifications for economic policy, land use controls, taxation, and environmental concerns.

To monitor whether the mandate for no net loss of wetlands is being followed, there needs to be accurate data regarding the spatial distribution of existing wetlands through a time interval. In other words, there is a need for a series of wetland maps, and there really is no other acceptable alternative. Armed with GIS and the appropriate series of digital maps, it would be relatively easy to evaluate the distribution of wetlands. For example, an important tool of modern GIS is the ability to measure the area of features. In fact once a wetland area is encoded into the system, its area is automatically calculated. In a digital world it is also easy to automatically compare wetland maps at different times to measure and to display changes.

Unfortunately, the fact that the tools exist to monitor wetland change does not mean that governments have been able to implement such systems. From a mapping science perspective several problems exist. The most fundamental issues are classification, delineation, resolution, and representation of wetland data. Although the FGDC is beginning to address the problems, we as a nation have not collectively agreed about what wetlands are, how to identify them, or how to encode them. The problems are difficult because wetlands themselves are ambiguous entities. Although wetlands are often included as a category of land cover, they actually are a condition of various categories of land cover (e.g., forested wetlands) and are commonly defined in terms of soil types. The wetland classification problem is exemplified in Table 6.1, which lists seven different wetland definitions. Although it is possible for the scientific and regulatory communities to reconcile differences in definitions, once we agree on what constitute wetlands, they still have to be identified on an acceptable mapping base. This task is difficult because the boundaries are usually transitional, without abrupt or definite edges. These fuzzy boundaries are rather elastic and can migrate with seasons or with climatic changes. The result are zones of varying transition.

Another issue involves the optimum way to represent wetland features in a digital spatial data base. For some programs and analytical needs, wetland data are stored as a cluster of uniform grid cells. Such raster systems often portray wetlands derived from digital image data captured by satellite or airborne photography. In most mapping and inventory systems used for planning and regulation, a wetland feature is represented as a homogeneous closed polygon. Within the existing programs there are several alternatives for the creation, maintenance, and distribution of a wetland data base. The possibilities include:

- a layer of polygonal areas that would replace the swamp symbols on the USGS 1:24,000 DLG versions of the topographic quadrangle maps;
- a series of polygons of the FWS 1:24,000 NWI maps;
- hydric soil polygons on the SCS soil surveys;
- a subset of wetland polygons that share boundaries with upland areas on a national land-use and land-cover data base;
- a digital layer of a proposed quarter-quad (1:12,000) orthophotography program (SCS, ASCS, and USGS);
- a digital layer on a federal, state, or local agency cooperative program such as the Maryland digital orthophoto quarter quad mapping and wetland inventory program; and
- a series of polygons on maps (digital or not) from the various state and local wetland mapping programs.

TABLE 6.1 Seven Examples of Wetland Definitions

Emergency Wetlands Resources Act of 1986 (PL99-645) The term "wetland" means land that has a predominance of hydric soils and that is inundated or saturated by surface or groundwater at a frequency and duration sufficient to support, and that under normal circumstances does support, a prevalence of hydrophytic vegetation typically adapted for life in saturated soil conditions.
Swampbuster Provision, Food Security Act of 1985 (PL99-198) The term "wetland," except when such term is part of the term "converted wetland," means land that has a predominance of hydric soils and that is inundated or saturated by surface or groundwater at a frequency and duration sufficient to support, and that under normal circumstances does support, a prevalence of hydrophytic vegetation typically adapted for life in saturated soil conditions.
U.S. Fish and Wildlife Service (Cowardin *et al.*, 1979, adopted 1980) Lands transitional between terrestrial and aquatic systems where the water table is usually at or near the surface or the land is covered by shallow water. For purposes of this classification, wetlands must have one or more of the following three attributes: (1) at least periodically, the land supports predominantly hydrophytes, (2) the substrate is predominantly undrained hydric soil, and (3) the substrate is nonsoil and is saturated with water or covered by shallow water at some time during the growing season of each year.
U.S. Environmental Protection Agency (40 CFR 230.3, Federal Register, 1980) *and the U.S. Army Corps of Engineers* (33 CFR 328.3, Federal Register, 1982) Those areas that are inundated or saturated by surface or groundwater at a frequency and duration sufficient to support, and that under normal circumstances do support, a prevalence of vergetation typically adapted for life in saturated soil conditions. Wetlands generally include swamps, marshes, bogs, and similar areas.
State of Wisconsin (NR 115.03 WAC) Those areas where water is at, near, or above the land surface long enough to be capable of supporting aquatic or hydrophytic vegetation, and which have soils indicative of wet conditions.

State of Connecticut (22a-38 Connecticut General Statutes) Wetlands means land, including submerged land, which consist(s) of any of the soil types designated as poorly drained, very poorly drained, alluvial or flood plain by the National Cooperative Soils Survey, as may be amended from time to time, by the Soil Conservation Service of the U.S. Department of Agriculture.
State of California (California Coastal Act of 1976, Section 30121) Lands within the coastal zone which may be covered periodically or permanently with shallow water and include saltwater marshes, freshwater marshes, open or closed brackish marshes, swamps, mudflats, and fens.

Mapping wetlands presents an interesting set of challenges. At the same time, our federal policies require that the problems be addressed. Therefore, the MSC examined the current roles of various institutional entities in the use and sharing of geographic information pertaining to the nation's wetlands. During the study, we found that policy issues dominated the consideration of wetlands; Appendix A treats these issues in additional depth. Wetlands illustrate how the scientific and resource planning and management communities go about the identification, classification, and delineation process in contrast to how a society goes about the difficult process of deciding what it is willing to regulate. Similar examples of national interest could be the distribution and condition of endangered species habitat or an assessment of biodiverse land areas.

EVALUATION PROCESS

Public Interest in Wetland Protection

One study (Dahl, 1990) estimated that 53 percent of the nation's original wetlands had been lost by the mid-1980s. Public interest in protecting and enhancing the remaining 103.3 million acres of wetlands emerged at the state and local level before the federal level. Congressional interest began in the late 1970s. In 1977 the Federal Water Pollution Control Act of 1972 was amended; it is commonly known as the "Clean Water Act." This act prohibits the dredging and filling of "jurisdictional" wetlands without a permit from the Corps of Engineers (COE). In 1985, and again in 1990, Congress expanded interest in wetlands by including economic incentives for farmers to eliminate "swampbusting" of wetlands in farmed areas. In 1986 the FWS was authorized to conduct a statistically based analysis of the status of trends of wetlands and produce NWI maps for the conterminous United States by 1998.

In 1988 the nations interest in wetland protection was amplified by President George Bush when he ran for election in support of the Conservation Foundation's recommendation of "no net loss" for the remaining wetlands. In parallel with this national focus and interest, 27 states now have some form of wetland protection regulations. In essence the debate over wetlands has now moved from a question of whether wetlands should be protected to a question of how much protection should be afforded the remaining wetland resource (Zinn and Copeland, 1992).

This question of how much or what subset of the overall wetland resource needs national attention and regulation, and a major debate is under way between wetland protection advocates and private land owners. The environmental and wetland advocates have concluded that "a major impediment in maintaining, enhancing, and restoring wetland resources is the lack of a *coordinated, consistent approach* among federal, state and local governments" (Zinn and Copeland, 1992). This lack of a coordinated and consistent inventory mapping and analysis capability is at the crux of public debate (see Appendix A). Without a reconciliation of the overlaps in these authorities and resultant data collection procedures, the ability to incorporate spatial data and information about the nation's wetlands into an NSDI remains problematical.

Lack of a Collective Perspective

It can be concluded that the need to bring forth a common view on the location, status, and trends of wetlands and the subset of those of national interest remains as the primary impediment to broad public and private support and the eventual criteria for spatial data about wetlands. For example, the owners of wetlands say "that protection efforts have gone too far" (Zinn and Copeland, 1992). A similar concern was reported in *USA Today* when President Bush was quoted "We ought to stay with our objective of 'no net loss'. . . but we don't want to overdefine what a wetland is" (Benedetto, 1992).

In essence, wetlands described and mapped as part of NWI and the Food Security Act (FSA) should be identical in definition and boundary delineation. Jurisdictional wetlands or that subset to be regulated by the COE and the EPA need to be nestable within the NWI and the FSA classification system. At present, jurisdictional wetlands are determined case by case by the COE, with veto responsibility assigned to the EPA. Because the COE conducts only case-by-case analysis, there are no maps available to alert those who might be affected by these jurisdictional wetlands. As stated by the Executive Director of the Association of Wetland Managers: " . . . in contrast [to the COE] virtually all state and local governments map wetlands as part of the regulatory process" and " . . . this lack of a consistent map base results in a federal program subject to varying interpretation by individual regulators." As a result, " . . . the regulatory process is difficult and time consuming. . . . Moreover, both delineating a wetland and applying for a permit are costly" (Kusler, 1992). (Niemann, 1992, estimated an annual cost of about $100 million.)

As in the previous paragraph, the MSC finds that without a reconciliation effort (among federal, state, and local regulatory activities) leading to a composite national view, incorporation of spatial data and information about the nation's wetlands into an NSDI remains problematical.

Impediments

Four potential impediments were identified that limit in various forms the implementation of a national wetland data and information resource. These impediments are: technical, legislative, institutional, and economic. These potential impediments were compared with 11 data collection and management recommendations prepared by the Interagency Wetlands Task Force (Table 6.2). The results of this analysis further demonstrate the difficulty of developing a national wetland information resource. The problems include inconsistent interpretations of wetlands between NWI and FSA maps even though the definitions are identical.

Other technical issues emerged, such as the technical specifications and data sources for wetland information. NWI delineations involve aerial photographic interpretations; FSA determinations involve the use of soil maps, aerial photographs, on-site evaluations, and more recently experimentation with satellite imagery. COE/EPA determinations of jurisdictional wetlands are handled case by case. This complexity of overlapping man-

TABLE 6.2 Impediments to a National Wetlands Spatial Data Infrastructure

Recommendations (Wetlands Inventory Report)	Technical Impediments	Legislative Impediments	Institutional Impediments	Economic Impediments
1. Complete NWI maps (FWS)	• Nonautomated product • No image backdrop (not essential but would enhance product usefulness) • Existing satellite resolution limits usefulness	• No authority to use FWS resources to automate (does not impede 1986 emergency mandate but restricts pace of automation and robustness)	• Other federal, state, and local agencies are duplicating wetland products	• Increased funding required
2. Integrate statistical analyses (FWS, SCS)		• Duplicative and nonintegrated efforts limit full value of wetland quantity and quality data collected	• EPA's EMAP is not included[a]	
3. Implement wetland change program for coastal wetlands (NOAA)	• Limited automated and compatible data sets restricts implementation	• Authority limited to wetlands associated with coastal areas	• Absence of other agency interests (e.g., FWS, SCS, and EPA) results in duplication	
4. Coordinate and integrate national wetland permit tracking system (COE-RAMS)	• Relationship to GRASS is not operational		• Long-term relationship with state permitting is unclear	• Increased funding required

Recommendations (Wetlands Inventory Report)	Technical Impediments	Legislative Impediments	Institutional Impediments	Economic Impediments
5. Develop automation standards for wetland information (FWS)	• Usefulness of SDTS is unclear	• No standard setting authority exists for automated wetland data	• Success of FGDC not assured	
6. Coordinate and integrate wetland mapping programs (FWS, SCS)	• Lack of compatibility between wetland data bases (1:24,000, NWI) in cartographic form with SCS (1:7,920) photographic base[b] • Interface with NWI maps and the SCS orthophoto program are not clear	• No legislative requirement to reconcile and integrate collection platforms, classification systems, and data sets		
7. Establish a National Wetlands Digital Data Base (FWS)	• Lack of automated data sets limits analysis • Inclusion of grade B data • Mixing data sources quickly • Maintaining attribute integrity	• Until recently, no authority to use FWS resources to automate		• Until recently, only available resources for automation were 100% user-pay dollars
8. Expand mapping and inventory systems to include functional value of wetlands (EPA, NOAA)	• Absence of reliable predictive models limits usefulness • Absence of low-cost high-resolution satellite imaging limits applicability		• Relationship with other mapping and statistical interests is duplicative (e.g., SCS, NWI)	

Recommendations (Wetlands Inventory Report)	Technical Impediments	Legislative Impediments	Institutional Impediments	Economic Impediments
9. Establish a national orthophoto program (1:12,000) (SCS/ASCS/USGS)	• Access policy to all image base products (hard and soft) by other federal, state, and local agencies not formulated	• No explicit authority exists to implement such a program	• Limited institutional interest	• Increased funding required
10. Coordinate large-scale digital wetland information (USGS)	• Actual mechanisms for assimilating large-scale data sets into small-scale data sets not yet developed	• No authority exists for agency responsibility	• Limited experience in assimilating data from non-federal sources	• No incentives for data sharing
11. Establish a national digital land cover program (USGS)	• Information technology has not been adapted to implement such a program • Classification system remains confused between land cover (reflective) and land use (activity) • Unit of resolution unresolved	• No explicit authority exists for such a program	• Relationship with proposed national orthophoto program is duplicative • Limited experience in assimilating data from non-federal sources	Increased funding required

[a] For example, others have concluded that . . . "For maximal usefulness (EPA's design) must be adopted by as many federal agencies as possible" (BEST/WSTB, 1992).

[b] Proposed orthophoto program offers a solution (see item 9). Also SCS is investigating satellite imaging systems for FSA wetland determinations.

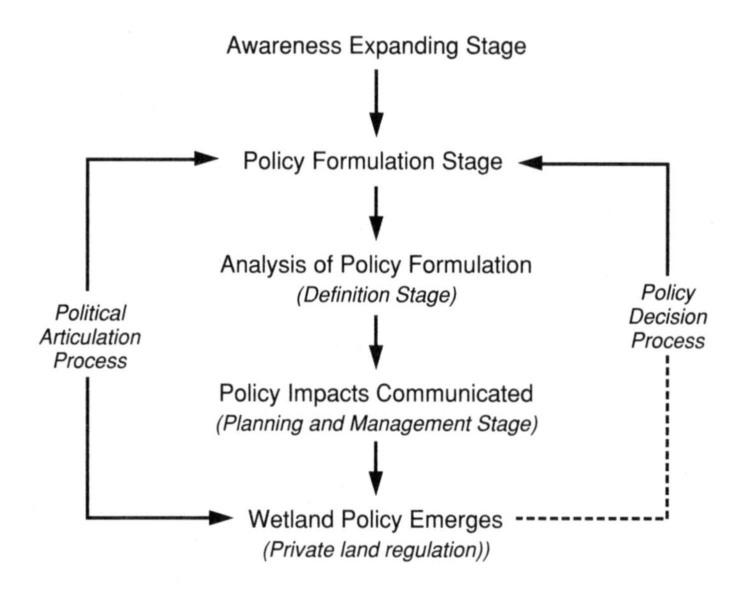

Figure 6.1 Evolution of wetland information technology diffusion.

dates and various analytical approaches compounds the data integration problem.

All of this complexity speaks to the need for a strategic approach for those responsible for administering wetlands policy. Elements of such a strategic approach could include the following:

- integrating and coordinating NWI and FSA wetland programs and nesting COE jurisdictional wetlands within NWI and FSA,
- developing standards for digital wetland information,
- fully implementing the multiagency national orthophoto program (both digital and hard copy products)
- integrating statistical analysis of the status and trends of the nation's wetlands, and
- integrating and coordinating the various ongoing state and local wetland mapping programs and nesting these classifications within the NWI.

The ability to implement such a strategy will require a concerted effort. The FGDC should consider facilitating such an effort.

Wetland Information Diffusion Model

To understand the current status of spatial data products that support decisions concerning the nation's wetlands, it is useful to examine the changing context of wetland information. Wetland data and information that initially were collected for descriptive and planning purposes are increasingly being used for regulatory purposes, which places different requirements on the information. The MSC presents these changing requirements in terms of information diffusion models. The description of an evolutionary model could serve those who will become responsible for such other natural phenomena of national attention. Figure 6.1 portrays an information diffusion model that explains and/or predicts the state and condition of wetland information. Five information stages were identified:

(1) awareness stage: the process of collecting and analyzing information to gain public support;
(2) policy formulation stage: the process of using information to gain legislative support and define intent;

(3) definitional formulation stage: the process of developing administrative rules that are dependent on reliable (repeatable) measures;
(4) planning, management, and analysis stage: the process of mapping wetland entities for both internal management and also to communicate the location and distribution of these regulated and non-regulated wetlands; and
(5) private land regulation stage: the process of imposing stated public policy on privately held lands.

What the model shows is that as the process moves into the last stage—regulating of private land—information requirements become more specific and demanding. Wetland entity mapping becomes integrated with property boundaries. Digital data collection and GIS technology and automated land records become important factors.

This evolution of wetland information diffusion is in a state of flux as portrayed in Figure 6.2. As this definitional process affects land development rights on privately held land, the public debate accelerates. This interaction, even though troublesome to the information community, in effect constitutes the current implementation process. As a result we have a difference between the policy of no net loss and the political process by which society and private land owners are determining what they are willing to endure.

Finally, in retrospect, if the information community had been in a position to implement a national wetland information resource before and in anticipation of the private regulation stage, it is likely that much of the present technical chaos, overlapping information collection mandates, and institutional inaction could have been substantially reduced. Therefore, the MSC finds that consistent and clear policy along with an understanding of the diffusion of information are required if the wetland components of the NSDI are to be strengthened,

CONCLUSION

We conclude that **the FGDC, through its Subcommittee on Wetlands, needs to reconcile the technical and institutional issues that impede our nation's ability to efficiently and effectively map, assess, monitor, and automate wetland information.** To accomplish this reconciliation, the FGDC needs to exercise its coordination authority that

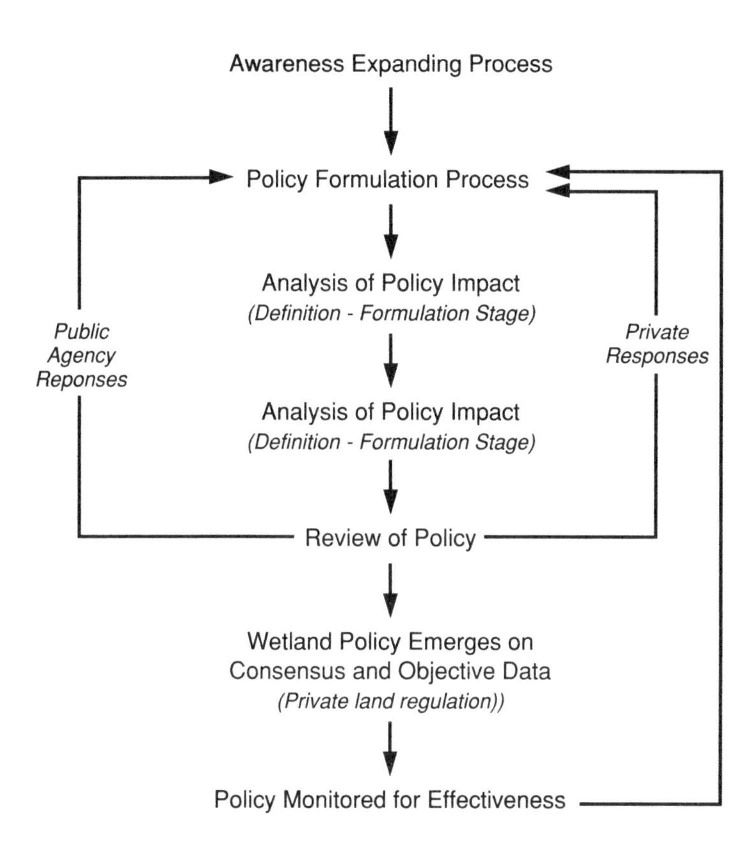

Figure 6.2 Proposed model for diffusion of wetland information.

resides in the OMB revised Circular A-16 and to develop and implement a wetland information diffusion model that is responsive to both policy and regulatory requirements. Many of the necessary reconciliation efforts are outlined in Appendix A.

REFERENCES

Benedetto, R. (1992). Bush pledges more for Forest—But other messages are mixed, *USA Today*, June 1, 1992.

BEST/WSTB (1992). *Review of EPA's Environmental Monitoring and Assessment Program (EMAP): Interim Report*, Board on Environmental Studies and Toxicology (BEST) and Water Science and Technology Board (WSTB), National Research Council, Washington, D.C., 25 pp.

Kusler, J., (1992). Wetland delineation: An issue of science or politics, *Environment 34* (2).

Niemann, B. J., Jr. (1992). Geographical Information Systems (GIS) Technology: Modernizing the Wetland Permitting Process, CRSS Architects, Inc., Houston Texas.

Want, W. L. (1991). *Law of Wetlands Regulation*, Clark Boardman Co., Ltd., New York.

Zinn, J., and C. Copeland (1992). Wetland Issues in the 102nd Congress, CRS Issue Brief, Congressional Research Service, The Library of Congress, Washington D.C.

7
SHARING OF SPATIAL DATA

With the increasing availability and use of geographic information systems, many governmental organizations, private companies, and academic researchers have the capacity to greatly expand the quantity, accuracy, and type of spatially referenced data available. With this capacity also comes the potential for substantial duplication of effort or the underutilization of valuable information that often has been created at considerable cost and effort.

RATIONALE FOR A SPATIAL DATA SHARING PROGRAM

Objectives

The principal objective of a spatial data sharing program is to increase benefits to society arising from the availability of spatial data. The benefits will accrue through the reduction of duplication of effort in collecting and maintaining of spatial data as well as through the increased use of this potentially valuable information. The exposure of these data to a wider community of users may also result in improvements in the quality of the data. This will eventually benefit the donor and other users.

The focus of a spatial data sharing program should be on increasing access to spatial data that are collected with the direct or indirect support of public funds. A spatial data sharing program should not displace the role of the private sector in providing value-added products and services associated with the utilization of this spatial data. It should, on the contrary, result in a rich environment for developing new business opportunities and enhancing economic growth.

Examples of Data Sharing Programs

The concept of sharing spatial data is not new. Examples can be found at federal, state, and local levels. The following two examples illustrate the types of efforts and benefits that can be derived from spatial data sharing. These examples cover two important types of baseline spatial data: geodetic detail and land parcel definition. There are many other kinds of spatial data as described elsewhere in this report, and there are enormous opportunities to reduce costs and increase user benefits through joint collection and sharing of spatial data. Data describing the location of a wide variety of phenomena can be shared: soil characteristics, wetlands, wildlife, hydrology, transportation systems, land use, and demographics, to name a few. All can be improved by organized joint efforts for their collection and distribution.

National Geodetic Reference System[1]

A number of incentives for data sharing and other forms of cooperation appear to have worked, and in some cases, very well. In 1980, for example, the NGS published standards for the submission of geodetic information to NGS. These volumes, known as the "Blue Book" (Federal Geodetic Coordinating Committee—FGCC, 1980, 1989), provided the specific descriptive information and formats for the mandatory and optional data elements for vertical (bench marks) and horizontal control data for inclusion in the National Geodetic Reference System (NGRS). The third volume of the trilogy, covering gravity control data, was published in 1983 (FGCC, 1983).

The Blue Book has evolved over the years in response to changes in surveying technology. For example Volume 1, *Horizontal Control Data*, was revised in January 1989 to include Global Positioning System (GPS) data submission and new formats for an improved, unified publication

---

[1] Information was supplied by Gary M. Young and Richard A. Yorczyk of the National Geodetic Survey in March 1992.

format for the descriptions that accompany published control-point values. As circumstances require, Blue Book requirements have been relaxed to accommodate unique situations. The USGS, in cooperation with the NGS, is currently converting much of its remaining third-order leveling data to computer-readable form so that these data can be incorporated into the North American Vertical Datum of 1988 (NAVD 88). This work is being accomplished with customized NGS software. The 10-year effort will eventually incorporate about 500,000 USGS bench marks into NAVD 88, thus vastly increasing the usefulness of the NGRS both in traditional leveling applications and through improved geoid modeling of regions that would otherwise be deficient.

This data sharing program works because the donors (private, county, state, and other federal organizations) want to ensure the accuracy of the points they observed (or had contractors observe) and earn NGS's stamp of approval as the nation's highest authority on geodetic control. It also provides the mechanism for the publication of officially sanctioned values, the national distribution of these values, and automatic updates of the data when future readjustments of NGRS are performed. Increasing the distribution of geodetic data in turn leads to an increased frequency of reuse of the control points by local and regional users, where each instance saves either private or public funds.

The NGS has received data from outside users for 65,000 horizontal control points since 1980. There has been a total of 36,000 km of geodetic leveling submitted to NGS by other organizations since 1980. A similar effort to gather private gravity data is currently under way. These gravity data will also be used to improve geoid height modeling, an essential requirement for accurate GPS-derived orthometric heights. The cost savings to NGS for these horizontal and vertical data conservatively can estimated at about $79.4 million (65,000 points × $1,000 per horizontal point = $65,000,000 and 36,000 km × $400/km of vertical data = $14,400,000).

North Carolina Land Records Management Program[2]

North Carolina has had a very active land records modernization effort since the inception of the Land Records Management Program

[2] Information supplied by Don Holloway, private consultant, on March 1992.

(LMRP) in 1977. The legislation provided for financial and technical assistance to local governments in the following areas: base maps, cadastral maps, a uniform system of parcel identifiers, and automation of land records.

One of the first major efforts was to develop standards for base and cadastral mapping and to arrive at a uniform parcel identifier. It was beneficial to the program that the Canadian project in Maritime Provinces of Canada was underway and the report from the National Research Council, *The Need for a Multipurpose Cadastre* (Committee on Geodesy, 1980), was published. These two major efforts were used in developing standards. Representatives from most state agencies and local government agencies that would be involved or benefit from the program participated in developing the standards.

Base maps (orthophoto and planimetric) were prepared on the State Plane Coordinate System, and a geo-coded parcel identification system was developed. Mapping on the State Plane Coordinate System was assisted by the fact that the North Carolina had a Geodetic Survey Office in place to establish and maintain a system of horizontal and vertical control monuments across the state. This office often gave priority to local governments in establishing additional control for mapping programs. Most counties elected to develop orthophoto base maps that were used to assist in the development of the cadastral maps. The photo image proved to be quite useful in establishing parcel boundary lines in this metes-and-bounds state. The photo image was also much better understood by the general public than was the planimetric map.

The fourth area of the program—automation—presented some interesting challenges in that many local governments had already begun automating their land records. It did not seem practical or feasible to standardize hardware systems across the state. Instead of standardizing computer systems, North Carolina opted to update, verify, and improve the data base as they developed the cadastral maps. This proved to be quite beneficial because many properties had not recently or ever been surveyed, and conflicts between property boundaries were discovered. In turn, property owners were notified of a potential problem and surveys ensued. The surveys were used to update the cadastral maps and data files.

North Carolina's financial assistance program of up to 50% of the total cost was important in the success of the program but not overriding. The state was willing to provide some seed money and establish the LRMP to provide assistance, which was more important in encouraging local governments in proceeding with this major effort than were the funds received. In presentations to local governments, the state always stressed that they should not proceed with a land records project because they would receive matching funds from the state, but because there was a need for more effective and efficient local government. Time has shown that North Carolina local governments have indeed benefitted from this effort.

During this process of assisting local governments in North Carolina, many side benefits were realized. Cooperation between the state and the counties began to evolve. Most of the counties had or were in the process of developing soil maps. This was a cooperative effort by the state, counties, and the SCS. As North Carolina moved into the digital mapping arena, duplication of effort in the soils mapping area was eliminated and sharing of soils files was effected.

In 1977 the state had established a Land Resource Information Service. This program was to establish and maintain a digital land based system for the state. Soils mapping was one layer that was needed and made possible because the counties were well on their way in developing soil maps. As the counties began to enter the digital environment, they agreed to provide the state copies of the soils mapping digital files, thus eliminating the cost to digitize the soils mapping. The state also agreed to furnish local governments copies of soil mapping files that had already been digitized, thus reducing the cost to local governments for this layer in the local data base. Forty-one of North Carolina's counties currently have digital mapping capabilities.

The LRMP also led to improved cooperation between state and federal agencies regarding land records. One legislative study committee learned that a federal agency and a county were having aerial photography flown on the same day at the same scale of the same area. Furthermore, the aircraft were in the same airport. This incident caused the legislative study committee to direct the LRMP to set up a meeting with all federal agencies that might be acquiring aerial photography in North Carolina to achieve greater cooperation between the state and the federal governments. This meeting did take place and proved to be most beneficial. Much greater cooperation was achieved, especially with the SCS and the USFS. This effort was greatly assisted by North Carolina's Geodetic Survey Office, which had a long-time association with the NGS.

A PROPOSED SPATIAL DATA SHARING PROGRAM

The MSC envisions a system that enables digital spatial data collected by nonfederal institutions (e.g., state and local governments and the private sector) to be integrated into the national spatial data coverage. The spatial data sharing program for the NSDI should provide access to digital geographic base data of known quality and currency through a limited number of access nodes linked to a fully decentralized communication system. We envision a system of spatial data servers, owned and operated by institutions authorized by the FGDC, to be the custodians of the data in question. These institutions are regarded as coproducers of the data as the data are produced according to previously agreed-upon standards with mechanisms in place to ensure their quality. To be successful the spatial data sharing program needs to have real benefits or incentives for both the donor and recipient of the data. A conceptual model of the program is shown in Figure 7.1; other details are given with Figure 7.2 and accompanying text.

Key Concepts

Types of Spatial Data

The NSDI consists of geographic base data and other spatial data. Base data are a primary geographic spatial reference that is produced to a recognized standard of accuracy and is subjected to certified quality assurance programs. Typically this is the type of data produced by federal and state agencies responsible for cartographic products (see Table 4.2). Other spatial data are available from a variety of producers whose standards for spatial accuracy may not be as rigorous. Data that are of less precise locational control often contain valuable supplementary information that cannot be found from base data sources; these data would include those representing a higher degree of currency or those of a thematic nature. These two types of spatial data can be treated somewhat differently within the NSDI as proposed in Table 7.1.

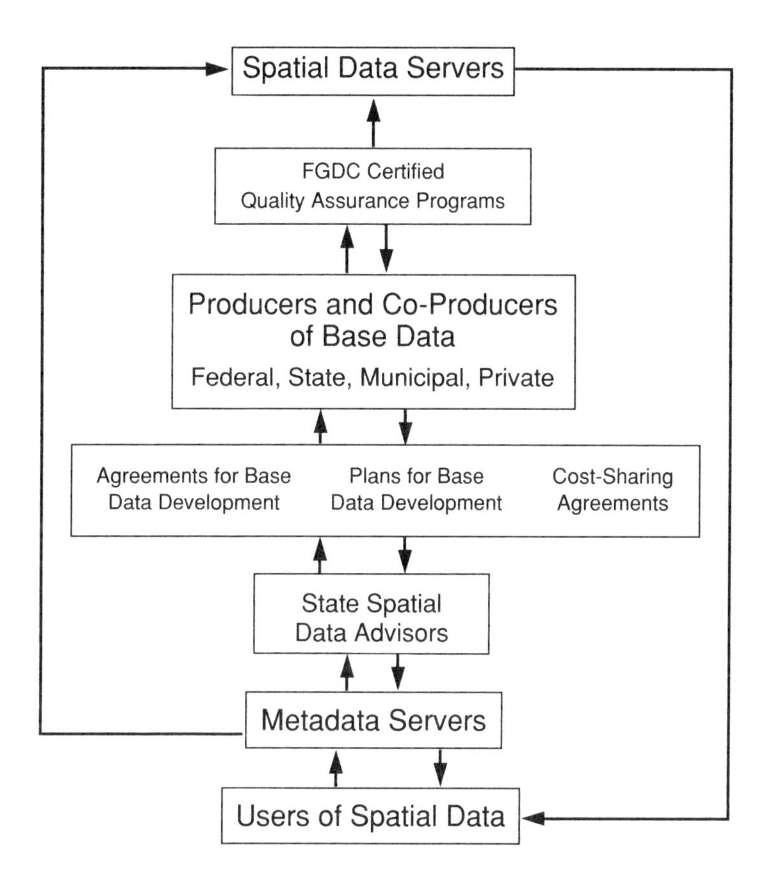

Figure 7.1 Components of a spatial data sharing program.

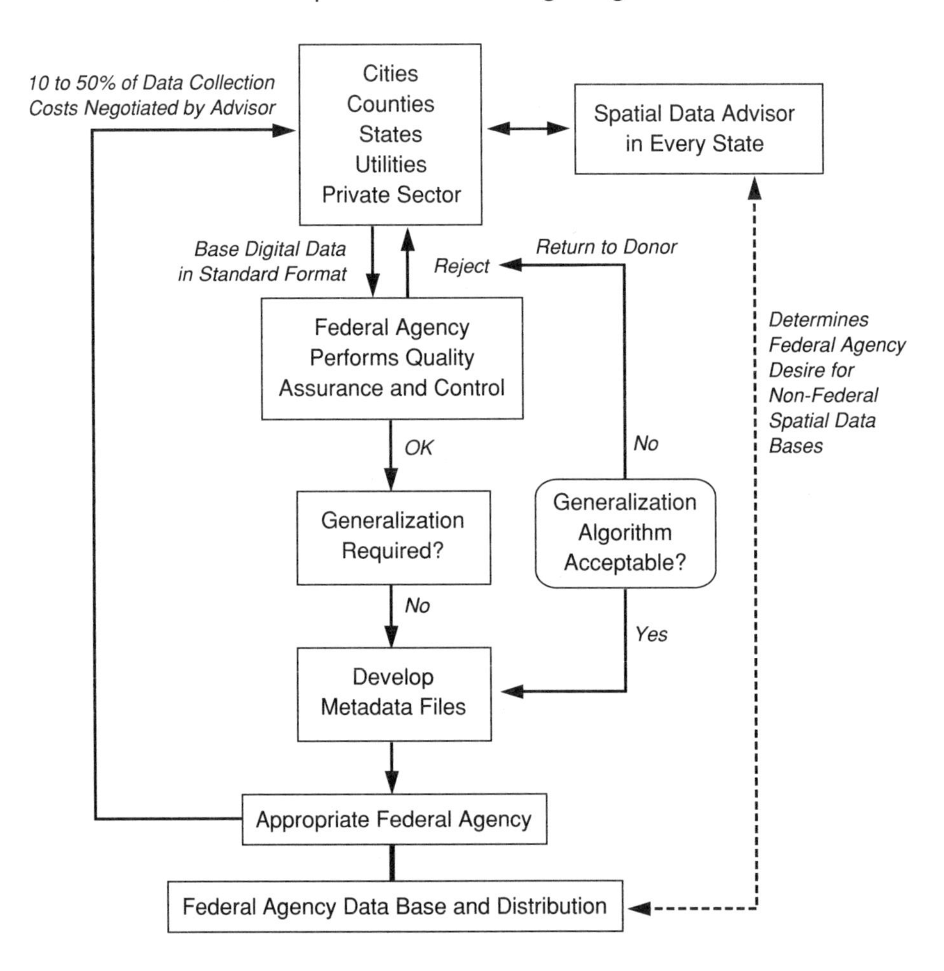

Figure 7.2 Representative procedures in a spatial data sharing program.

TABLE 7.1 Treatment of Base Data and Other Spatial Data Within the Proposed Program

Characteristics	Base Data	Other Spatial Data
Accuracy standards, quality assurance, and independent certification	Must comply	Statement of estimated accuracy
Cost sharing between co-producers	Available/ encouraged	Encouraged
Include metadata descriptor	Required	Required
Comply with Spatial Data Transfer Standard	Must comply	Must comply
Metadata servers	One or a few	Many
Spatial data servers	Possibly many	Many

Quality Assurance for Base Data

In the NSDI accuracy standards will need to be set for base data. We anticipate that these standards will continue to be established by the federal agencies with specific responsibility for different geographic features (as per OMB Circular A-16). These standards should be coordinated and disseminated as NSDI standards through the FGDC.

Metadata

An important aspect of data access and retrieval is knowledge of their existence, contents, and fitness for an application. This knowledge is referred to as metadata, or data about information. Metadata describe the content, ancestry and source, quality, data base schema, and accuracy of data. Metadata support data sharing by providing information on many aspects of spatial data, each aspect having meaning in particular application contexts. Metadata that describe data base contents include data dictionaries and definitions, attribute ranges, and data types. The origin or ancstry of data is critical for ascertaining the validity and suitability of data.

The metadata file descriptors are an important part of the SDTS. The development of the metadata standards and protocols will enable the creation of an easily accessible, networked data base that can be searched, preferably on-line, by the users seeking particular types of information. These metadata bases may also be used in the future to determine data gaps or duplication in the public national data base.

NSDI Spatial Data Sharing Program

The proposed spatial data sharing program (Figure 7.1) in many respects represents a combination of elements of both the Geodetic Advisor Program of the National Geodetic Survey System and the North Carolina LRMP (discussed above). Figure 7.2 shows how such a sharing program might be implemented.

State-level spatial data advisors (similar to the existing state-level geodetic advisor) would determine what base data being collected (or planned to be collected) within their state might be suitable for incorporation into the National Geographic Data System (NGDS). An advisor would contact the appropriate federal agency (presumably the lead agency on the FGDC for the data category of concern) to determine whether a given data set might be included in the federal data sets. If the federal agency agrees to consider these additional data, a plan would be developed for providing work sharing or partial reimbursement of costs for the collection of such data. Elements of this approach currently exist on an ad hoc arrangement between some federal agencies and states; these often include either work-share or cost-share agreements. If the data collection is planned, the federal agency would work with the nonfederal entity to build into the data collection program the appropriate standards, accuracy, and quality control of the resulting data. If the nonfederal data sets of interest currently exist, the federal agency would evaluate the potentially donated data sets.

Such data sets would be provided by those who collected the data (potential donors) to the appropriate federal agency in a standard format for a quality assurance and quality control (QA/QC) analysis. If the data set fails to meet the established QA/QC criteria, the data would be returned to the donor. If the data set meets the QA/QC criteria, the next question would be if generalization of the data (e.g., from a large scale to a smaller standard scale) is necessary. If generalization is needed, the appropriateness of the algorithms would be determined. If these were not acceptable, the data would be returned to the donor. If the generalization algorithms are acceptable or if generalization is not necessary, the federal agency would develop the appropriate metadata files for the data set and incorporate the data set into the national base.

The incentives for donors to submit their data to be considered in the national base would be threefold. First, a portion of the costs of data collection would be rebated to the collector, the amount being coordinated and negotiated by the state-level advisor with the federal agency; if the data are not yet collected, then work- or cost-sharing arrangements might be struck. Second, the donors would have the assurance that the data they collected (or had contractors collect) meet accepted national standards and have been subjected to an independent QA/QC analysis. Third, the program provides the mechanism for the broad national distribution of these data and other data in addition to updates of the data when future revisions are made from other donor sources.

A number of questions remain unanswered, all of which will require further analysis in developing a workable data sharing system. These include (1) How should the state data advisors be funded? (2) Can a single advisor handle the flow of data from the respective state? (3) What scales should be allowed? (4) What agencies should be responsible for the QA/QC? Should it be the lead agency designated by FGDC? (5) Should that same agency be responsible for developing the metadata files? (6) Who returns a portion of the data collection costs to the data donor? Is this a function of the agency that has stewardship of a given data type or layer within a broad National Geographic Data System (FGDC, 1991)?

Guidelines for System Implementation

Who Donates?

Currently there are many federal government agencies providing their data to the public on a timely basis and at a nominal cost. The quantity and quality of information that can be obtained through federal government agencies is extremely rich and has been used in countless applications. For example, the availability of TIGER geography, national base mapping (USGS), and other national and international spatial data (through NOAA, SCS, DMA, and others) has provided many individual users and organizations with an opportunity to leverage their own activities and avoid unnecessarily duplicating effort.

Clearly all federal government agencies should be participants in a spatial data sharing program. Similarly, state agencies and local governments involved in the collection of base data (see Table 4.2) could and should be active participants. For example, local governments are most directly involved with new street names and address ranges.

Apart from government agencies funded directly by tax revenues and private companies or academic institutions that may gather spatial data with the use of direct public funding, there should be no requirement that compels any organization to become a data donor. Public-spirited private companies and academic institutions may voluntarily elect to participate in a spatial data sharing program as a result of their own belief in the value and importance of sharing certain data with the rest of society; the exchange of such data for cost-equivalent access to other shared data would be another incentive for private companies.

What Data Are Donated?

A spatial data sharing program should place special emphasis on the collection and dissemination of primary or base data. What do we mean by base data? Are TIGER data, for example, a base data set or are they partly derived (from DLG) data? These are thorny issues that will arise with any sharing program, and it may not be prudent to attempt to exclude any type of data from the program. Rather, the emphasis may need to be placed on ensuring that the ancestry of any spatial data is unambiguously documented so that the user is fully aware of its origins and limitations. In some instances there may be some justification to disseminate spatial data that are clearly secondary (i.e., derived from a primary data source). The justification for dissemination of these data may be that the process of transforming the primary data into secondary data is very time consuming and that most work is with the data in the secondary form.

There are, however, certain types of base data that help form the structural backdrop for a large number of currently collected spatial data. In defining the base data for a NSDI, the information that is basic to one user may be selective to another and vice versa. However, in this report we refer to base data as the information required to establish a basic reference to the Earth's surface. To this basic data set can be added features, attributes, and other intrinsic information. However, it defines a clear framework to reference data from many sources. In defining base data it quickly becomes apparent that accuracy and detail of content vary by use and scale of operation. Therefore, we have selected four levels to be used in reference to base data (see Table 4.2). Finally, for brevity and understanding we avoided a detailed set of specifications and approached the task by relating it to map scales, accuracies, and content. We realize that certain digital data are multiscale, but accuracies and content are determined by scale.

A very important dimension of the spatial data sharing program would be the emphasis placed on adhering to standards. The program should endorse one or two standard data transfer formats as the official currency of trade in the program (e.g., SDTS, VPF). There will be some questions here as to whether large quantities of information available in previously used government standards should be or need to be converted to any new standard before becoming available through the spatial data sharing program. This may require a transition strategy. In the long term, the spatial data sharing program should encourage the use of a limited number of standards. The standards should be publicly accessible and not proprietary. GIS software companies will eventually respond through the normal forces of the market and enable their software to both read and write to the standards required by the spatial data sharing program.

In addition to a file format standard, there needs to be a metadata file describing the content, ancestry, quality, and accuracy of the data being made available. Such information is proposed to be part of SDTS. Consideration should be given to embedding the metadata file descriptor in the programs' selected transfer standard. The development of the metadata standard will enable the creation of data bases that can be accessed and searched by users seeking particular types of information. These metadata bases may also be used to determine where there may be data gaps or duplication in the public national data base. Before a few years ago, strong arguments could have been made for a centralized catalog or metadata directory. There are developing computer networks, through such programs as the National Research and Education Network (NREN) of the High Performance Computing and Communications Program, that will enable the establishment of a distributed metadata directory system (assuming that standards and protocols are invoked). These networks can also provide a mechanism for data distribution; however, the desirability of this can be dependent on the data transfer rates and the slowest component on the network.

The spatial data sharing program may provide access to other components (other than data) of the infrastructure. Some of these components might include people, software, and facilities. The spatial data sharing program should be designed to accommodate such components in the future. An exception to this might be the availability of data translator software, for example, which converts spatial data in a file format not supported under the spatial data sharing program into an acceptable format. This would enable organizations with very large inventories of spatial data in unacceptable format to allow users to convert this information with their own time and effort.

What Are the Incentives and Requirements to Donate and Distribute?

The incentive to donate information to the spatial data sharing program will in part have to be driven by a public sense of responsibility and a recognition that in many instances the beneficiaries of the program will be the data donors themselves. In turn the donors will be able to reduce their costs by avoiding the collection of redundant and duplicate information. Additionally, the donors might receive a rebate to help offset the costs associated with the data collection or assistance through work- or cost-share programs. An important incentive for those participating in sharing of base data might be an independent assessment of data quality.

Federal agencies have little incentive to incur the incremental expense of adhering to standards and coordinating activities with other agencies when undertaking a spatial data generation program. There is currently no means by which the potential users of those data could share in the additional incurred costs, and therefore in periods of tight budgets agencies tend to do the minimum that is necessary to perform their basic mission.

In this environment it may be helpful if the federal government adopt a general statement of policy that spatial data created by any federal agency be made available in accordance with standards. Other agencies requesting the data should be prepared to bear the additional costs of adherence to standards. To ensure compliance this policy should be made a part of each agency's annual appropriations. The benefit of the standardization of data to all governmental agencies—federal, state, and local—and to the private sector is such that this incremental cost will be recovered to the federal treasury over time as direct savings in government programs and in increased efficiency in the private sector. This assertion is borne out in the many studies done on benefits of the use of GIS (see Chapter 3).

Additionally, the FGDC should assure the OMB that proposed federal programs that will gather significant quantities of spatial data will not duplicate data that already exist. The OMB, in their budget-examining role, should rely on the FGDC's assurance that proposed federal programs for spatial data are nonduplicative.

The federal government could substantially influence participation of state and local governments by making it a requirement of its numerous grant programs that if spatial data are collected, these data are made available through the spatial data sharing program.

What Are the Rules for Usage?

Access to spatial data under the sharing program should be available to everyone. In some instances agencies may wish to restrict the usage of spatial data to particular target groups, but it can be argued that the administrative effort required to establish the credentials or appropriate conditions for exclusive access are too difficult to administer and add unnecessary costs. The essence of the spatial data sharing program should be to disseminate information that by its very nature is in the public domain.

In this respect there should also be no restriction for usage. Private companies should be able to freely access this information to build value-added products or services. Although no restrictions would apply, it may be an advantage of the program to ensure that organizations, private or public, that use data under the spatial data sharing program must acknowledge the donor (by adopting the metadata file descriptor) so that the consumer can be fully informed of its origins.

The FGDC needs to further investigate the legal liability for the quality and accuracy of donated spatial data. This is an extremely important dimension that needs to be addressed promptly so that it does not become a constraint to the success of the program.

Who Supports the Users?

The availability of spatial data will result in consumer questions about the data. Questions may be associated with the technical format of the data or relate to the data content. The increased availability could result in an initial additional burden being placed on many organizations to answer questions about their data and data-collection activities. This is an inevitable consequence of increased public scrutiny and awareness of spatial data. Some organizations participating in the program will find that they need to place more effort in improving data dictionaries and other documentation regarding their data products. Although initially there may be some difficult adjustments, in the long term a spatial data sharing program will lead to an improvement in the value and accuracy of donor data-gathering activities. In addition, for data sharing to succeed beyond the original use of the data, a mechanism for continued data maintenance must be built into any support mechanism.

In some instances, agencies may be reluctant to participate in the program because of their concern that the increased public awareness of their spatial data may disrupt their ongoing programs. Although the MSC has some sympathy for the management of these organizations, the public's right to know and use these data must be the paramount consideration.

MECHANISMS FOR IMPLEMENTING A SPATIAL DATA SHARING PROGRAM

The proposed spatial data sharing program must do more than just disseminate spatial data collected by federal agencies. The richness and utility of the program is substantially enhanced by having participation of donors from state and local governments, academic, and the private sector. Unfortunately, there is no current mechanism in place for such participation in the operation of a program of this nature. A challenge in establishing the spatial data sharing program is, therefore, to determine whether this can occur without a formal organizational structure or, if necessary, what the optimum structure would be. An additional and formidable challenge is how the spatial data sharing program should be funded to be successful.

The FGDC is the federal program with the objectives and intent most comparable with this program, albeit without funds. The MSC believes that at this time the FGDC should assume the initial leadership role to embrace the broader scope of this proposed spatial data sharing program.

The FGDC should establish a data sharing committee with the objective of providing the policy making and leadership to launch, maintain, and operate the proposed program. Membership of the committee would consist of representatives from the federal community (FGDC) as well as appointed/invited representatives from state and local governments as well as academia and the private sector.

The data sharing committee would not be responsible for any operational programs other than establishing policies, monitoring and evaluating the performance of the program, and communicating the existence and value of the program. The principal policy areas that the committee should address include the following:

- data standards policy,
- depository policy,
- distribution policy, and
- cost sharing policy.

The data standards policy responsibilities would include defining from a technical viewpoint the proposed metadata model for describing and categorizing spatial data donated under the program. Under this umbrella the committee could also select those federal data standards that would be accepted under the program (SDTS, TIGER, etc.). Finally, and at a high level, the committee may also wish to endorse the dissemination of certain federal conversion utilities that exchange data from one federal standard to another.

The depository policy role of the committee should be to establish guidelines for how federal agencies should participate in and comply with the program. Guidelines or recommended regulations could also be drafted for how federally funded cost-shared projects and programs ought to make spatially referenced data available. Guidelines and legal conditions and liability limitations for organizations and agencies volunteering to deposit data under the program could also be documented.

The distribution policy for the spatial data sharing program would be defined for both the metadata base as well as the specific geographic data files. Costs associated with fulfilling the distribution of the data should be borne by the end user (consistent with current federal policy on data distribution). Guidelines for technical user support should also be specified under the distribution policy.

The operation of the spatial data sharing program will require some financial resources, but the bulk of the operational costs should be borne by the donors and recipients of data from the program. The overhead associated with the operation of the committee and the maintenance and distribution of the metadata base should be funded initially through the FGDC.

The first few years of the implementation of such a program will undoubtedly expose many issues and difficulties, some of which may not be easy to resolve or reconcile. These difficulties should not be permitted to distract from the central theme of this initiative that a cooperative environment can greatly benefit the nation. Although some initial financial support from federal agencies may be required to initiate this program, the MSC believes that the benefits of this investment will greatly outweigh the initial costs.

SPATIAL DATA CATALOGS

As mentioned previously, one of the needs for a robust NSDI is a mechanism for identifying the full range of spatial data collected, where the data are stored, who controls access to the data, the data content, the metadata, and the areas of data coverage. Spatial data catalogs provide an important component of the NSDI, one that can be established by using distributed computer networks.

Distributed Data Catalogs

Software can provide on-line search capabilities of catalogs of spatial and other data that are resident on a computer connected to a network of other computers. Such a capability, however, requires that the data catalogs be accessible in a standard protocol on the servers. The goal is to have information searches coherent across different services.

One such program is WAIS, which is a public-domain software program developed jointly by Dow Jones News Service, Peat Marwick, Apple Computer, Thinking Machines, Inc., and others. WAIS uses the Z39.50 protocol and the Internet computer network of networks to scan, search, and often access existing data bases. The Z39.50 standard or protocol that allows WAIS and other software to search distributed data bases is a product of the National Information Standards Organization (NISO), accredited to the American National Standards Institute (ANSI). Z39.50 is fully compatible with the NISO standard for library catalogs (Z39.2), originally promulgated by the Library of Congress and known as MARC (MAchine Readable Cataloguing), and has a corresponding

International Standards Organization (ISO) standard. Computer-to-computer interchanges, whether components of the Z39.50 protocol or of the content being delivered, are precisely represented in a standard computer language known as Abstract Syntax Notation (ASN.1).

WAIS implements Z39.50 in a client/server mode of computer interaction. In a typical search for textual information, the client software prompts the user to select which information sources to include in the search and to enter a search request. Once the search request is entered, the client software converts the search words to the standard information retrieval protocol (Z39.50) and presents the search request in turn to each server associated with a selected source. The server software takes the words and matches them to the contents of all documents in each selected source. The client software receives search results from all of the servers and presents to the user a list of all document or data base titles found. When requested by the user, the client software requests the server to pass the full contents of the document or metadata file and presents the document to the user.

The use of such software evokes the experience of using a library. A library user may begin by consulting a card catalogue or index or by asking a reference librarian. At this point, the user is searching for documents based on a few key words (e.g., subject, title) or names (e.g., author). The user reviews the documents found and may note other key words or names that could lead to additional relevant documents. A feedback situation develops as the user modifies subsequent searches based on results found in prior searches. Ideally, the user stops searching when all the most relevant documents are found.

Information Servers

Information servers using WAIS can be registered to a Directory of Servers currently maintained on Internet by Thinking Machines. The registration entry includes text information about the contents of the sources reachable through the server, and this information is itself indexed for searching. Also listed is information that will be used by the client software to contact the server (e.g., TCP/IP node name) as well as information on what and how to pay charges for use of the server if it is not free. Indexing of text to create an information source is fairly rapid, a 30 megabyte file was indexed in about 20 minutes on a Data General Aviion (Eliot Christian, USGS, personal communication, 1992).

Any server capable of responding to Z39.50 information retrieval requests can be an information server. Information servers can be local (on the workstation or local area network) as well as remote (accessible now via TCP/IP, in the future via X.25 networks or asynchronous dial-up). WAIS does not require any central coordination unless the server is to be advertised through the Directory of Servers. In fact, an information server registered to the Directory of Servers can itself act as a subordinate directory of servers administered locally. By describing sources under various directories of servers, it is possible to organize the sources in whatever relationships make sense and yet allow users to search as many sources as desired.

One feature of WAIS that is allowed but not required by the Z39.50 protocol is that the client/server interaction is stateless: at the application level each request from the client to the server is a separate process that is not associated with any previous request. The server does not maintain information about the client between requests. This feature is very significant for situations in which a user may want to search hundreds of sources on dozens of servers at a sitting.

Information Sources

Information servers provide access to the information sources placed on them. These sources are compilations that may include a variety of formats. Such formats are known as document types, although information need not be textual. Although all Z39.50 clients and servers support search and retrieval of textual information, support for other document types that may have been registered in Z39.50 is negotiated when the client initiates a relation with a server.

When sources are created, defining the document types allows the server to use the appropriate translation between the specific query format of the source and the Z39.50 protocol. The public domain WAIS package includes assistance in creating information sources and provides indexing software for several common document types consisting of text, graphics, and bibliographic references in MARC. Source code in the C programming language is provided for adding other document types. If access to other data structures is required, the server interface routines are also designed to be customized. A typical customization would be to use search requests to access a relational data base using Structured Query Language.

Applications

The USGS is using WAIS to enhance the Earth Science Data Directory (ESDD). The ESDD is maintained as a source for references to earth science data, including many at the state level and a comprehensive list of data holdings relevant for arctic research. WAIS is especially appropriate for that application, because the ESDD user community ranges from local citizenry to international global change researchers. The ability of WAIS to place the ESDD in the context of other USGS and external information sources is especially powerful for these users. The USGS intends to publish (and maintain) in the WAIS Directory of Servers, a subordinate directory of servers focused on earth science data and information.

The USGS is adding features required for ESDD (phrase searching, location searching, and key word searching within fields), which can be accommodated within WAIS. The USGS is also including the ability for a user of the client software to drop from a WAIS session into an automated log-in to existing data systems, such as the Global Land Information System (GLIS). With this approach, users of the USGS/WAIS client software would be able to access any Z39.50 server but would have additional capabilities when accessing one of the USGS servers. The USGS is also using WAIS to access a clearinghouse of USGS spatial data holdings.

The ESDD interfaces with the interagency Global Change Master Directory, a single interagency source for references to key global change data. The global change data management community is considering WAIS as an adjunct to the Global Change Master Directory. As a data directory tool, it is possible to rapidly correlate the Global Change Master Directory to existing data directories relevant to global change research. For example, NOAA has a directory with about 25,000 data set references and the Inter-university Consortium for Political and Social Research has another directory referencing about 28,000 data sets.

The ability of WAIS to handle different information sources through a single user interface makes it possible for researchers to explore publications and data sets concurrently. The federal research libraries involved in global change research (primarily NASA, NOAA, USGS, and USDA) are very interested in the potential for WAIS to bridge between the data and information worlds. Also, WAIS is seen as a useful way to connect textual information with a data system. For example, when a user is researching an existing data set, it would be useful to provide immediate access to all of the associated documentation about that data set. The use of WAIS, the associated documentation could extend beyond the data set itself to include publications that reference the data set or engineering specifications of the instruments used.

REFERENCES

Committee on Geodesy (1980). *Need for a Multipurpose Cadastre*, National Research Council, National Academy Press, Washington, D.C., 112 pp.

FGCC (1980). *Input Formats and Specifications of the National Geodetic Survey Data Base: Volume II—Vertical Control Data*, Federal Geodetic Coordinating Committee, National Geodetic Survey, Rockville, Maryland.

FGCC (1983). *Input Formats and Specifications of the National Geodetic Survey Data Base: Volume III—Gravity Control Data*, Federal Geodetic Coordinating Committee, National Geodetic Survey, Rockville, Maryland.

FGCC (1989). *Input Formats and Specifications of the National Geodetic Survey Data Base: Volume I—Horizontal Control Data* (revised 1989), Federal Geodetic Coordinating Committee, National Geodetic Survey, Rockville, Maryland.

FGDC (1991). *A National Geographic Information Resource: The Spatial Foundation of the Information-Based Society,* Federal Geographic Data Committee, First Annual Report to the Director of OMB, 10 pp. plus 41 pp. of appendixes.

8
CONCLUSIONS

The MSC concludes that the NSDI needs to be improved if we are to succeed as a highly competitive nation. A great deal can be done to improve the infrastructure and recommendations are presented in Chapter 9 to do that.

In doing so, we admit that we have taken a federal perspective, caused by the fact that federal agencies fund the committee and hence get attention, and by the background of the committee members. We plan to address state, local, and private aspects more fully in a following report.

As previously stated, spatial information is in a period of transition between traditional paper records—mostly maps—and sophisticated digital data bases. These data bases will be increasingly important in a major paradigm shift in the next decade. Chapter 4 addresses issues associated with another step toward this new paradigm. In this chapter the MSC characterizes its vision of the new paradigm.

Briefly, this new paradigm enables the customer to specify, over a computer network, what kind of product is wanted rather than the present situation where sales of an existing product are advertised. Goods will be ordered and paid for electronically. A user will be able to drive down the street with a house you might purchase while sitting at a computer. Some are presently using services such as Prodigy© or Compuserve© to specify, for example, a subset of the *Wall Street Journal* to look at each evening in contrast to looking at the entire newspaper. Given these functions and many others that we cannot envision, we begin by describing the principles of and proposed enhancements to the NSDI.

IMPROVING THE NSDI

The Principles

There are four critical principles that need to guide the development of the NSDI (after Dertouzos, 1991): availability, ease of use, flexibility, and a foundation for other activities.

Availability

The NSDI is a national strategy and is not designed to serve the interests of one level of government, one sector in society, or one geographical area. The data should be available through public networks that have maximum user capacity or other media (such as CD-ROM).

Ease of Use

Weiser (1991) notes that "the most profound technologies are those that disappear. They weave themselves into the fabric of everyday life until they are indistinguishable from it." New generations of computer appliances and standard software will help to take the use of spatial data from the hands of the technical specialists. Accessing spatial data should become as easy as turning on a light switch: the complexity of networks, standards, and data base structures should be transparent to the user.

Flexibility

The NSDI cannot be dependent on the technology, data, or organizational structures of today; it must be able to anticipate and manage growth. It must cater to the needs of many different types of users and incorporate many types of data. If the potential dissemination of spatial data is to be realized, for example, then communication networks must be capable of handling a wide range of transmission speeds to accommodate everything from simple text to four dimensional animation.

Foundation for Other Activities

The infrastructure is not an end in itself but the means of realizing the value of spatial information. Its purpose is to foster and not to control new applications, services, and industries.

The Components

The components of the NSDI include the data bases (and metadata) and their sources, the spatial data networks and their users, the technology, the institutional arrangements, and the policies and standards required to coordinate all of the various parts (see Figure 3.1). The objective here is to sketch the NSDI components as background for the issues involved and development of a master plan.

Data Bases, Metadata, and Sources

Although many types of spatial data are now in digital format, conversion of the base data (e.g., topography and land tenure) has been costly and slow. In a national strategy, there is a need to identify priority data bases on the basis of a solid understanding of user requirements and the NSDI potential. Or as Dertouzos (1991) defines the problem:

> ". . .we, the designers and users of this information infrastructure, bear a serious responsibility: we must understand the value and role of information so that we may better channel our technological miracles into useful rather than frivolous, if not dangerous, directions."

Because the potential sources and users of the data will range from federal agencies to private local organizations, the priorities need to be determined in as wide an environment as possible and not controlled by one particular sector.

An important subcomponent will be the metadata and standard tools for geographical referencing. The development of directories incorporating these metadata is an important feature of a robust NSDI because it is necessary for all other applications. It is an achievable task in the short term, and nondigital or incomplete databases can be incorporated. Special tools for displaying and accessing geographically referenced information, such as electronic atlases, also will be important in the NSDI.

Data Networks

The networks are the highways linking data bases and users in the NSDI. This is not a vision of one coaxial cable running across America; in fact the networks are referred to as clouds, indicating a complex configuration of communication media, data transmission schemes (e.g., packet switching), traffic control mechanisms, and gateways connecting different subnetworks (Cerf, 1991). For the NSDI, the objectives will include designing a network configuration that will provide flexibility (e.g., varying transmission speeds for different data formats), maximum national coverage and user accessibility, and the security and reliability required for various types of enquiries.

The NSDI network will be composed of many subsystems, including dedicated telephone lines, local area networks, integrated-services networks (ISDNs), and new communication systems designed to handle the increasing volumes of heterogeneous data. For example, Broadband ISDN is currently being investigated internationally by telephone companies as a means to provide a common network for all communication services and information, in contrast to the special networks now in use for different services such as voice, data, and video (Cerf, 1991).

Technology

The NSDI provides a conduit for data. There must be intelligent connections at each end, optimizing the management of the data bases at the source and maximizing the potential application of the data by users. In between there will be, for example, an array of common servers providing standardized information services to users, gateways and query languages regulating the data flow, and interfaces translating the data from one format at the source to another at the user's terminal.

Negroponte (1991) emphasizes that "the added intelligence at each node and at the ends of the network are what make the system work." One important trend will be new types of terminals (e.g., writing tablets and wall screens) that will be integrated into our lifestyles (Weiser, 1991). Other trends include open system architecture (Tapscott, 1991); a greater use of artificial intelligence in managing data, networks, and applications; and the use of groupware and easily transportable application software. As Tesler (1991) points out "software . . . will change more than any other element in the computing paradigm."

Institutional Arrangements

Whether the NSDI will be effective will depend more on institutional arrangements than on technology. Without coordination of many different organizations and without leadership at the federal level, the NSDI will be reduced to a series of projects of limited value and lifespan. As has been demonstrated in previous ventures in spatial data sharing, organizational cooperation is the critical ingredient that will make or break the best devised plans.

The first steps will be to develop a common vision and to begin to build an organizational structure that can manage the construction and maintenance of the NSDI. Only then will it be possible to address the numerous legal, social, and financial issues that form the NSDI institutional environment.

Policies and Standards

Policies and standards are the heart of the NSDI and have an impact on all other components. Standards are the rules and common conventions that will allow data to pass from source to user. They will affect all levels of the infrastructure from technology and communication protocols to data content and use. Open system architecture will reduce the issues related to data exchange standards, but there are many other areas that will require agreement. In many cases standards will be set outside the NSDI by, for example, system vendors and network managers. Other standards, such as geographic referencing frameworks for data integration, must be specifically addressed in a spatial data context (e.g., Lee and McLaughlin, 1991).

Policies, whether formal or informal, establish the environment within which the NSDI will be developed and managed; they define the constraints and goals and somewhat delineate the means by which the goals will be achieved. There are numerous issues that need to be resolved, including the need for financial and political commitment. Some policies will be at the data and technical level and can largely be controlled by the organizations involved. Issues such as privacy and accountability need to be addressed within broader legal and political environments (e.g., Gore, 1991; Kozub, 1991).

Users

The users are given separate mention here because, in the flurry of designing and implementing the NSDI, it will be easy to become immersed in the technology, concepts, and data. Users probably will be the most mentioned group and yet actually the least considered. Unless there are user communities and unless the NSDI is changed to meet their needs (i.e., enables them to do something new or something they already do more easily or more cost efficiently), then the rationale for the NSDI falters (e.g., Weiser, 1991). That does not mean that all potential user groups or applications need to be identified, but it does mean that users be considered as part of the total infrastructure and that real rather than theoretical requirements are met.

A NEW STRATEGY

Given these principles and our findings, the MSC proposes a strategy for an enhanced NSDI. Table 8.1 outlines some of the key activities that should be included in such a plan.

There are several key activities that must be included in the NSDI. They are described briefly below but are not presented in a chronological or stage-by-stage manner because most activities will be both continuous and concurrent.

TABLE 8.1 Elements of a Strategy

- Obtain and maintain national commitment
- Evaluate requirements, constraints, and opportunities
- Evaluate the current status
- Determine priorities
- Develop coordination and organizational structures
- Assign roles and responsibilities
- Develop standards and policies
- Develop and monitor projects
- Identify and resolve the issues

Obtain and Maintain National Commitment

A continuing but strong commitment is needed at the outset. Commitment in terms of financing, policies, and resources must be obtained first at the federal level because it is a national strategy. However, it also will be important to convince policy makers and governments at other levels of the need for investment in the NSDI in the early organizational stages. Ongoing commitment will be as crucial as the initial enthusiasm. Although this will depend on economic, political, and other factors beyond the control of the NSDI organizations, one objective of the NSDI should be to show tangible benefits in the early stages of development to demonstrate the feasibility and the value of the initiative.

Evaluate Requirements, Constraints, and Opportunities

The determination of the various requirements, constraints, and opportunities for an NSDI will involve on-going research with major investments in the initial stages. The research must cover local to global considerations and be concerned not only with data and technology requirements but also with the management and institutional needs and constraints.

Research needs to be done to evaluate the current status of inventories of data bases and sources, networks, and services together with their specific characteristics. In addition to providing a basis for determining priorities, if standardized and kept current, this inventory could be used to develop electronic directories. Once again, the evaluation should go beyond the technical and data level and look at organizations and other institutional concerns.

Determine Priorities

From the evaluations of the requirements and current status, the next step will be to determine priorities. These will include priority data bases to be included in the NSDI, priority issues that need to be resolved, and priority services that should be made available.

Develop Coordination Mechanisms and Organizational Structures

A wide range of options for coordinating bodies and structures to initiate and manage the NSDI development should be investigated. Included in this review should be examples from other countries that have national spatial data organizations, such as Australia, Canada, and Germany, as well as examples from other nationally based activities in the United States. The organizational structures should be able to evolve to accommodate the changing priorities and needs of the NSDI development.

Assign Roles and Responsibilities

The development of the NSDI will involve not only the public sector (e.g., departments and committees) and major private organizations (e.g., private network managers) but also academia and special organizations. The latter may, for example, represent special citizen interests, small businesses, vendors, and users such as environmental groups. One of the initial tasks within the organizational arrangements will be to define the role that all such parties will play and their specific responsibilities within the NSDI. At this stage the issue of private sector involvement must be addressed at a policy and management level.

Develop Standards and Policies

A continuation of current research and coordination initiatives, particularly at the federal level, this stage will be crucial in determining the data bases that can be part of the infrastructure and in developing the information marketplace.

Develop and Monitor Projects

Building the NSDI will involve projects designed for the NSDI specifically (e.g., a prototype spatial data network) and undertaken by external groups for other purposes (e.g., distributed networks in municipal or state governments). Internal projects will need coordination by the NSDI organizations; mechanisms will also be required to identify and monitor external projects to determine their impact on the NSDI.

Identify and Resolve Issues

A coordinated research strategy should be designed to identify and propose solutions to the policy, management, and technical issues. At the technical level, for example, one issue that should be addressed is the future requirements for standards (i.e., beyond SDTS). Research funds and resources will have to be identified at an early stage. The NSDI organizations will also be responsible for ensuring that the issues are resolved through policy, legislation, regulation, agreements, or other means.

REFERENCES

Cerf, V. G. (1991). Networks, *Scientific American 265*(3), 72-81.

Dertouzos, M. L. (1991). Communications, computers, and networks, *Scientific American 265*(3), 62-69.

Gore, A. (1991). *Scientific American 265*(3).

Kozub, N. E. (1991). An exercise in strategic planning: Development of data policy in the information age, *Urban and Regional Information Systems Association 1991 Annual Conference Proceedings*, vol. 4, 39-49.

Lee, Y. C., and J. D. McLaughlin (1991). Distributed land information networks: Database management issues, *Canadian Institute of Surveying and Mapping Journal 45*(3), 231-238.

Negroponte, N. P. (1991). Products and services for computer networks, *Scientific American 265*(3), 106-113.

Tapscott, D. (in association with DMR Group) (1991). Open systems: Managing the transition, *Business Week*, October 14, 131-162.

Tesler, L. G. (1991). Networked computing in the 1990s, *Scientific American 265*(3), 86-93.

Weiser, M. (1991). The computer for the 21st century, *Scientific American 265*(3), 94-104.

9
RECOMMENDATIONS

1. Effective national policies, strategies, and organizational structures need to be established at the federal level for the integration of national spatial data collection, use, and distribution.

Experience has proven that certain data are required by every critical national program, for example, environmental cleanup, urban development, disaster relief, health care and disease control, industrial development, and transportation control and expansion. Lack of quality spatial data remains an impediment to industry and government efforts to address these critical national issues. Therefore, national policy goals should denote the concept of an NSDI as well as a strategy, and a time table for implementation must be set in motion.

The functions of an information organization, wherever assigned, should be (1) coordination of data collection activities, that is, to ensure quality data in standard formats and eliminate costly duplication of data collection; (2) development and coordination of standards; (3) assurance that data are easily accessible to the public through a catalog (or protocols for a distributed, networked catalog) of the data, including metadata or data about data; and (4) definition of incentives for a data donor/sharing program.

When these considerations are viewed in the context of the information age, maintaining national competitiveness in certain technologies, and the oft-stated desire to reduce bureaucracy, the need for a high level government-wide focus on spatial data is obvious. In our opinion this focus can best be accomplished by a government-wide reorganization. However, this has been attempted at periodic intervals and has failed. Pressures in the marketplace and public demands for better utilization of our informational resources make this option worthy of reconsideration.

Other options include an independent information agency, but that too is probably impractical. Other organizational possibilities are government-owned corporations such as the Tennessee Valley Authority or a council. These are also difficult to establish. The most practical solution, although not our choice in the best of all worlds, would be to strengthen or augment programs in existing agencies or departments.

Although most of the civilian mapping authority has traditionally been associated with agencies within the Department of the Interior, we assert that an equally acceptable site for a spatial information authority would be within the Department of Commerce. The Department of Commerce should be considered as a possible location for the NSDI authority both because of the important contributions that the private sector will make to the infrastructure and because of the implications for economic development and international competitiveness associated with the NSDI. Within Commerce, the Bureau of the Census has major responsibilities in collecting and analyzing a wide variety of information that is spatially referenced and created the TIGER spatial data set, which was designed for the 1990 census. The National Institute of Standards and Technology (NIST) has been involved in the SDTS process and other standard setting activities; standards are an important part of their mandate. Also within Commerce, the NOAA has significant data collection functions housed within the National Ocean Service and large data centers within the National Environmental Satellite and Data and Information Services. The vast majority of spatial data that are collected within the United States have either economic or environmental uses, which also makes the Department of Commerce a logical home for an information-based authority. *In either case, this program deserves priority consideration at the cabinet level and requires the backing of legislation specifying required funds and objectives.*

2. The Federal Geographic Data Committee (FGDC), which operates under the aegis of the Office of Management and Budget (OMB), should continue to be the working body of the agencies to coordinate the interagency program as defined in OMB Circular A-16. However, strengthening the charter and programs of the FGDC are needed to

- **expand the development and speed the creation and implementation of standards (content, quality, performance, and exchange), procedures, and specifications for spatially referenced digital data, and**

- **create a series of incentives, particularly among federal agencies, that would maximize the sharing of spatial data and minimize the redundancy of spatial data collection.**

Although the Spatial Data Transfer Standard (SDTS) has become a FIPS (FIPS-173), the fact remains that there is no common data exchange mechanism in the federal government. Further effort required includes the development of a variety of implementations of the SDTS and its extension and generalization to accommodate those essential data elements desired by data users. It is increasingly important to spatial data users to be able to communicate among various vendors' systems, and this interchange is possible only through the use of sophisticated and complete data transfer standards. The leadership and participation of the FGDC will continue to be important in establishing and implementing standards throughout the federal government and underscore the criticality of sharing and integrating digital data to meet national needs.

The incentive to share spatial data for public use will in part need to be driven by a public sense of responsibility and a recognition that in many instances the beneficiaries of sharing will be those who allow their data to be shared because they in turn will be able to reduce their costs by avoiding the collection of redundant and duplicate information. Government agencies have little incentive to incur the incremental expense of adhering to standards and coordinating such activities with other agencies when undertaking a spatial data generation program. There needs to be incentives to promote sharing and coordination of spatial data activities.

In fulfillment of the FGDC's mission, the following specific actions within this recommendation include:

- The FGDC should be empowered to help ensure that proposals to generate spatial data do not duplicate data in existing data bases that could be processed to accepted standards to satisfy the proposed purpose.
- The OMB should work with the FGDC to increase the budget planning allowance for each FGDC member agency as a means for obtaining funding to ensure compliance with national spatial data standards and minimize duplication of effort, implement the SDTS federal profile consistently across the federal government, and ensure SDTS library services to all users (federal, state, local, and private).
- To execute its mission effectively, the FGDC should include staff functions for lead agencies for a specific data type; assist with cataloging

of federal spatial data bases; verify that data bases are maintained and updated; provide liaison with state, local, and private agencies that either generate or use data of a specific category; and begin establishing the National Geographic Data System.

- The FGDC should consider establishing an additional subcommittee on national cooperative digital land bases. The subcommittee should be charged with developing within a 2-year period a plan and program for the development and maintenance of national cooperative digital land bases. Representation on the subcommittee should include the diverse interests found within all levels of government.
- The FGDC should evaluate the potential benefits of amalgamation and augmentation of the three independent street centerline spatial data base (SCSD) efforts now under way within the federal government. Substantial redundancies can be avoided by combining the contents of TIGER, ZIP+4, and the transportation layer of the 1:24,000 DLGs into one integrated data base and establishing ways for other organizations to attach specific attributes to a centralized resource under constant maintenance.
- An institutional and/or organizational structure should be developed to focus and encourage the many local initiatives for improvements in TIGER, ZIP+4, and DLG data sets. A mechanism needs to be established to accept enhancements to the SCSD while maintaining specified accuracy standards.
- The FGDC, through its Subcommittee on Wetlands, needs to reconcile the definitional and technical issues that impede our nation's ability to efficiently and effectively map, assess, monitor, and automate wetlands information.

3. **Procedures should be established to foster ready access to information describing spatial data available within government and the private sector through existing networks, thereby providing on-line access by the public in the form of directories and catalogs.** Information dissemination by federal agencies is included in the proposed revisions to OMB Circular A-130 (*Federal Register*, *57*(83), p. 18300), which states: "Agencies shall maintain and implement a management system for all information products which shall, at a minimum . . . (c) Establish and maintain inventories of all agency information products; (d) Develop such other aids to locating information products disseminated by the agency, including catalogs and directories . . .." Participation of nonfederal

governmental agencies and the private sector would be voluntary (see Chapter 7 for discussion of a spatial data sharing program).

An increasingly important class of information is metadata, which describe or annotate in some way the characteristics of the data. Examples of metadata might be how to access data, their ancestry, location, quality, validity, and accuracy. Directories and catalogs may represent both metadata and information. Such directories will become one of the gateway mechanisms for information access.

Consideration should be given to embedding the metadata file descriptor in the transfer standard. The development of the metadata standard will enable the creation of a distributed data base that can be accessed and searched by users seeking particular types of information. This metadata base may also be used to determine where there may be data gaps or duplication in any public national data base.

Capabilities for accessing and searching multiple data directories in diverse locations over computer networks (e.g., NREN or Internet) are being developed. These capabilities enable the stewardship of federal spatial data bases to remain with the originating (or lead) agency as defined in OMB Circular A-16 while still making the metadata, and ultimately the data per se, available. These network systems mean that one agency does not have to build a centralized catalog system while continually trying to coax other agencies to submit their metadata in a timely manner. With the adoption of standards and protocols (e.g., Z39.50) and acceptance of the distributive network concept, the first steps to a cooperative spatial data infrastructure will be in place.

In this environment it may be helpful if the federal government adopt a general statement of policy that spatial data created by any federal agency be made available in accordance with standards. To ensure compliance, this policy should be made a part of each agency's annual appropriations. The benefit of the standardization of data among all governmental agencies (federal, state, and local) and the private sector is such that this incremental cost will be recovered to the federal treasury over time as direct savings on government programs and in increased efficiency in the private realm.

4. A spatial data sharing program should be established to enrich national spatial data coverage, minimize redundant data collection at all levels, and create new opportunities for the use of spatial data throughout the nation. Specific funding and budgetary cross-cutting responsibilities of federal agencies should be identified by the OMB,

and the FGDC should coordinate the cross-cutting aspects of the program.

The spatial data sharing program should enable digital spatial data collected by nonfederal institutions (e.g., state and local governments and the private sector) to be integrated into the national spatial data coverage. The program would reduce data collection redundancies (and thus costs) throughout the nation as a whole while improving the currency of the data. To be successful, the spatial data sharing program needs to have real benefits and incentives for both the donors and recipients of the data.

As envisioned, incentives for donors to submit their data to be considered in the national base would be threefold. First, a portion of the costs of data collection might be rebated to the collector, the amount would be coordinated and negotiated by the state-level advisor with the federal agency; if the data are not yet collected, then work- or cost-sharing arrangements might be effected. Second, the donors would have the assurance that the accuracy of the data they collected (or had contractors collect) meets the accepted national standards and has been subjected to an independent QA/QC analysis. Third, it also provides the mechanism for the broad national distribution of these and other data in addition to automatic updates of the data when future revisions are made from other donor sources. Crucial to the development of incentives is an acceptance of the concept that a more efficient, more robust, more useful NSDI can exist and that it should replace the highly fragmented, highly redundant, often frustratingly inadequate, ad hoc infrastructure that exists today.

APPENDIX A
SPATIAL DATA AND WETLANDS

INTRODUCTION

What, typically, could be done better or more efficiently if the content, accuracy, organization, and control of spatial data were different? As spatial data concerning wetlands are collected in several federal agencies, state and local governments, and private institutions, the MSC believed that a study focused on wetlands would provide an example of the needs and challenges facing the development and implementation of a robust NSDI. With respect to wetland data, a number of questions arise. For example, should a digital version of the National Wetlands Inventory (NWI) be used to replace the wetland symbols on the 1:24,000 USGS series? Is it to be a distributed layer to be used as a graphic or digital overlay to the 1:24,000 NWI series or integrated within the new federally proposed 1:12,000 orthophoto program (SCS, ASCS, and USGS), or is it to be a distributed responsibility where each federal, state, and local agency provides its part in a coordinated and integrated form? An example of the latter is the joint effort between Maryland's digitial orthophoto quarter quad (1:12,000) mapping and wetland inventory program and the FWS's NWI program. These are important questions in need of answers.

To assist in clarifying these questions, this appendix focuses on the current roles of various institutional entities in the use and sharing of geographic information pertaining to the nation's wetlands. It comments on the impediments that exist that prevent these groups from acquiring knowledge, sharing data, making decisions, or performing the duties expected of them that depend on the timely availability and easy access to an organized body of geographic information about wetlands.

Initially the MSC wanted to study the technical problems associated with an uncoordinated NSDI relating to wetland data. However, when we began to study the problem, an even larger set of problems emerged. Therefore this case study has the following three goals:

- to determine the feasibility of establishing a national system that delineates and records the conditions of all regulated and nonregulated U.S. wetlands;
- to describe the impediments (if any) that limit this nation from delineating and recording the condition of its regulated and nonregulated wetlands; and
- to consider the extent to which these impediments are indicative of other natural phenomena of national consequence requiring delineation, monitoring, and eventual regulation.

Wetlands were chosen as an example because they reflect environmental and physical phenomena that need to be measured, depicted, and analyzed differently than discrete objects such as building footprints or street centerlines and associated street addresses. Wetlands were also chosen because they are of national concern and interest. They are indicative of how our nation goes about administering and managing natural resources. Wetlands also illustrate how the scientific community goes about the identification, classification, and delineation process in contrast to how a society goes about the difficult process of deciding on the subset it is willing to regulate. Similar examples of national interest could be the geographic distribution and condition of endangered species habitat or species ranges or a national assessment and monitoring of biodiverse land areas. If these two examples were to become issues of national interest what could be learned from this nation's attempts to map, monitor, and regulate its wetlands?

This appendix introduces the issues regarding wetlands that make them possible to measure scientifically but difficult to regulate as a natural resource. Wetlands are excellent examples of informational needs about other natural phenomena. Next are described the technical, legislative, institutional, and economic impediments that limit the ability to assess and monitor the state and condition of its wetlands. This appendix also provides a conceptual information diffusion model that attempts to explain the issues that restrict the diffusion of wetland information. It concludes with a summary with recommendations.

THE NATURE OF WETLANDS

What Are Wetlands?

The term wetlands applies to a variety of low-lying areas where the water table is at or near the surface of the land, soils are saturated or covered by water during parts of the year, and there is a predominance of hydrophytic plants (CEQ, 1989). In a more practical sense, the term wetland is a misnomer: many are dry at times; some are dry twice a day, for example, coastal wetlands that are flooded, inundated, and influenced by daily tides. Wetlands include many different types of environments: tidal marshes, swamp forests, peat bogs, prairie potholes, wet meadows, and similar transitional areas between aquatic and terrestrial environments.

Wetlands were long considered insect-ridden, unattractive, and dangerous areas. Recently this outlook has changed dramatically because the vital ecological roles that wetlands serve have been documented and thus have in this century begun to be recognized as important places with a rich and exciting variety of plant and animal life (Niering, 1986).

What is the Value of Wetlands?

Wetlands are among the most biologically productive ecosystems in the world. Net primary production of plants in salt marshes and freshwater wetlands (Figure A.1) rivals that of tropical rain forests and the most productive agricultural land (CEQ, 1989). For example, many types of animals depend on wetlands for at least part of their life cycle (e.g., it has been estimated that more than 50 percent of the saltwater fish and shellfish are dependent on wetlands). Of the 10 to 20 million waterfowl that nest in the conterminous 48 states, 50 percent or more reproduce in the prairie pothole wetlands of the Midwest (CEQ, 1989).

The wetlands of the United States are also important for other reasons. They produce oxygen and play a significant role in converting atmospheric nitrogen, for they naturally trap and remove nutrients and sediments and help maintain or improve water quality (Ducks Unlimited, 1992). Wetlands associated with estuaries, rivers, and streams, as well as some isolated wetlands and lakes, provide flood protection by slowing and storing floodwaters and reducing flood peaks. Wetlands anchor shorelines and provide erosion control (CEQ, 1989).

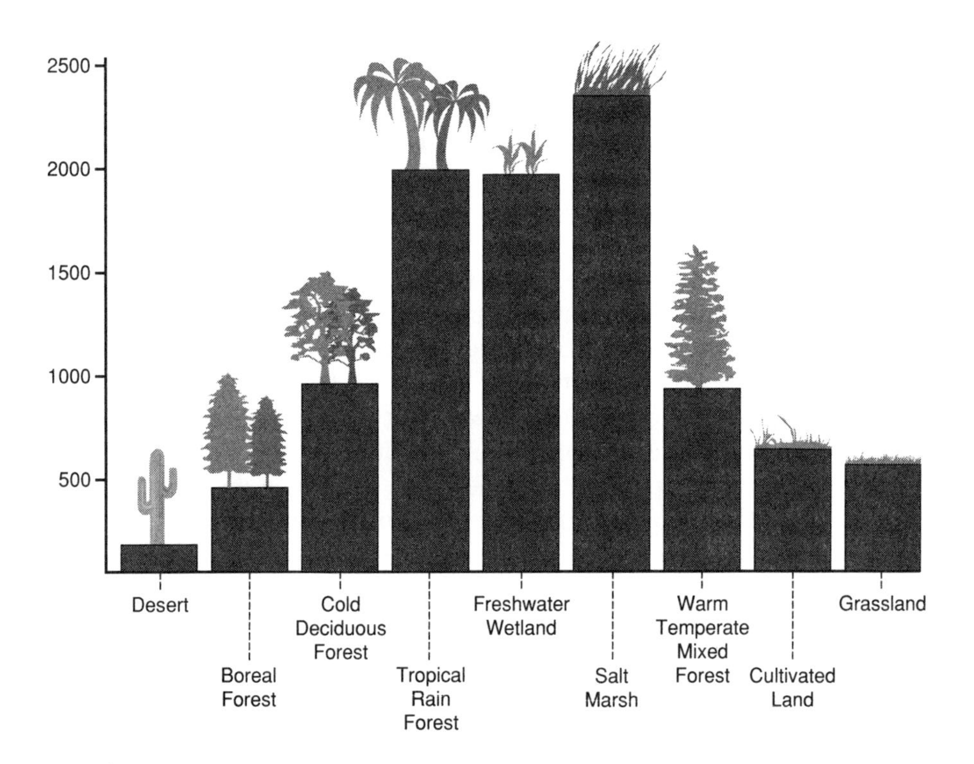

FIGURE A.1. Net Primary Productivity of Selected Ecosystems ($g/m^2/yr$) (from Tiner 1984).

Wetlands also provide many economic and social benefits to the nation. Fishing, waterfowl hunting, and traditional gathering of food, such as wild rice, are among the contemporary uses. Wetlands occur in every state in the nation (Figure A.2) but exist in a variety of sizes, shapes, and types as a result of regional differences in climate, vegetation, soils, and hydrology. Also, and very important to many, is that wetlands are some of the last remaining wilderness areas in the nation (CEQ, 1989).

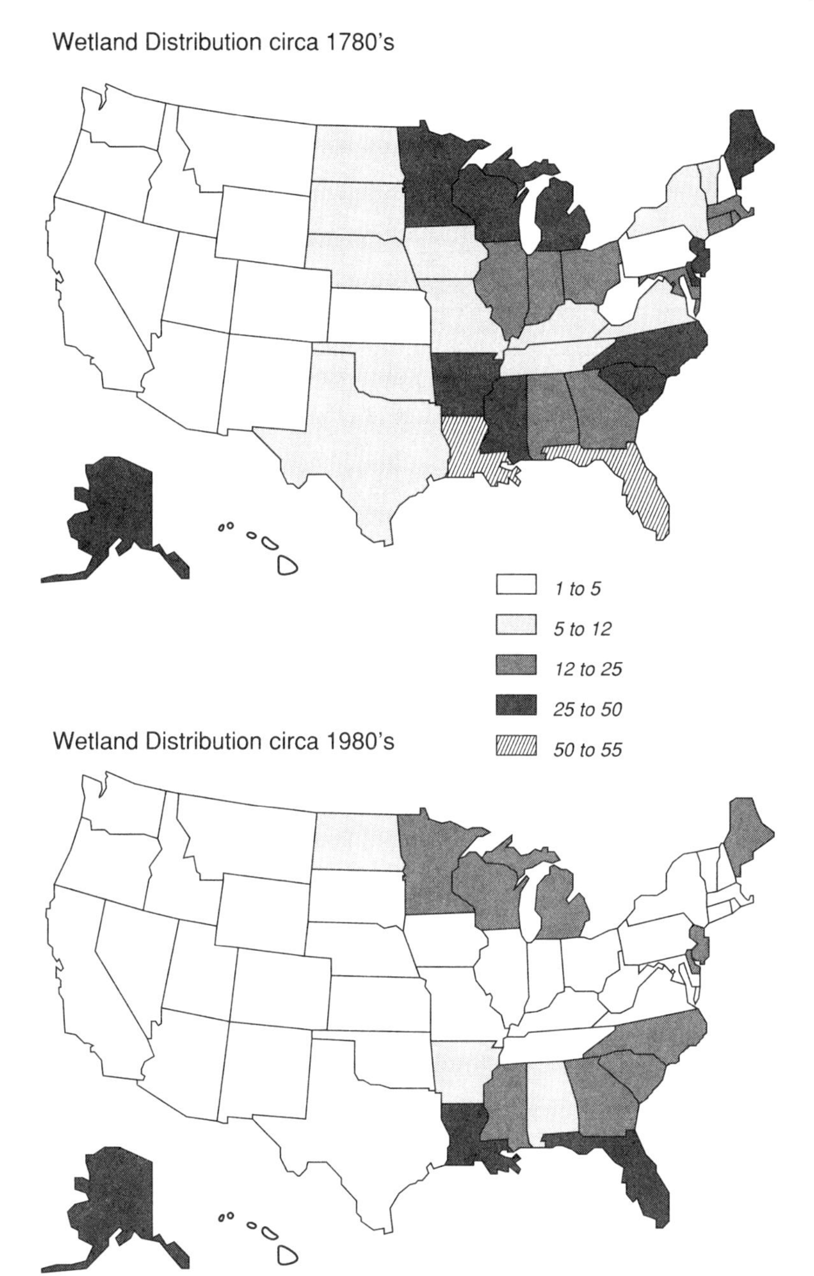

FIGURE A.2 (a) Wetland distribution circa 1780s; (b) wetland distribution circa 1980s (from Dahl, 1990).

TABLE A.1. Examples of Wetland Losses in Various States (after Dahl, 1990).

State or Region	Wetlands 1780s (acres)	Wetlands 1980s (acres)	Percentage of Wetlands Lost
Iowa's natural marshes	4,000,000	421,900	89
California	5,000,000	450,000	91
Nebraska's Rain-water Basin	94,000	8,460	91
Mississippi alluvial plain	24,000,000	5,200,000	78
Michigan	11,200,000	5,583,400	50
North Dakota	4,927,500	2,490,000	51
Minnesota	15,070,000	8,700,000	42
Louisiana's forested wetlands	11,300,000	5,635,000	50
Connecticut's coastal marshes	30,000	15,000	50
North Carolina's pocosins	2,500,000	1,503,000*	40

* Only 695,000 acres of Pocosins remain undisturbed; the rest are partially drained, developed, or planned for development.

STATE AND CONDITION OF WETLANDS

Wetlands are generally classified as estuarine or freshwater systems (CEQ, 1989). In the mid-1980s, there were an estimated 103.3 million acres of wetlands in the conterminous United States, most of which (about 75 percent) were private (Dahl and Johnson, 1991).

Wetlands account for roughly 5 percent of the total land surface cover in the conterminous 48 states. The amount of wetlands in the conterminous United States when settlement occurred in the early seventeenth century is estimated to have been 215 million acres (Dahl, 1990). On the basis of this

TABLE A.2. Examples of Wetland Loss Rates (after Tiner, 1984).

State or Region	Loss Rate (acres/year)
Lower Mississippi alluvial plain	165,000
Louisiana's forested wetlands	87,000
North Carolina's pocosins	43,500
Prairie pothole region	33,000
Louisiana's coastal marshes	25,000
Great Lakes basin	20,000
Wisconsin	20,000
Michigan	6,500
Kentucky	3,600
New Jersey's coastal marshes	3,084 50*
Palm Beach County, Florida	3,055
Maryland's coastal wetlands	1,000 20*
New York's estuarine marshes	740
Delaware's coastal marshes	444 20*

* Loss rate after passage of state coastal wetland protection laws.

figure (which is considered a reliable estimate of the original wetland area), 53 percent of the original wetlands was lost by the mid-1980s (Dahl,1990). Table A.1 shows a selected state and regional view of wetland losses.

The reasons for these wetland losses are many and varied and both natural and human. However, most of the wetland losses are attributable to human activities. Agricultural activities were responsible for 54 percent and urban development accounted for 5 percent of the total wetland loss (Dahl, 1990).

Wetland losses have affected certain wetland types more than others (Table A.2). By the mid-1950s it was estimated that more than 50 percent of the wetlands in the prairie pothole region had been lost since settlement. Most of this reduction was caused by agricultural conversion (CEQ, 1989). Between the mid-1970s and mid-1980s net acreage of freshwater marsh loss had stabilized (Dahl, 1991).

Wetland losses also occurred in estuarine vegetated wetlands (estuarine intertidal vegetated). Of the net loss of 372,000 acres between the mid-1950s and 1970, most occurred in estuary marshes along the Gulf Coast in Louisiana, Texas, and Florida. Urban development and conversion to open water habitat were responsible for most of these losses (CEQ, 1989).

PROTECTION OF WETLANDS

State interest in protection of wetlands began in the east. For example, Massachusetts' regulation of wetlands includes coastal wetlands, freshwater wetlands, swamps, wet meadows, marshes and bogs, and a 100-foot buffer protection zone. In Massachusetts judicial interest began in 1965 (Commissioner of Natural Resources *v.* Volpe and Co.); 27 states now have some type of wetland law (Want, 1991).

Federal interest in the protection of the nation's wetlands began in the mid-1970s. Various conservation groups and the scientific community began convincing federal agencies and Congress of their value in preventing floods, filtering waters, and providing critical wildlife habitat. The Federal Water Pollution Control Act of 1972 (often called the Clean Water Act) was amended in 1977 to prohibit the discharge of dredge or fill material into wetlands without a permit. Federal wetland law is still the backbone of wetland protection (Want, 1991). Expanding on this federal interest, President Bush in 1988 adopted a platform supporting the goal that there should be no net loss of the remaining wetlands (103.3 million acres) (Conservation Foundation, 1988; Seligmann and Hager, 1991).

This position, however, has raised the significant question of what subset of all wetlands should be included. This issue of whether all wetlands or subsets should be regulated is a major reason why it is so difficult to bring about a national system for wetland information. Wetlands occur in all 50 states and vary in type, size, and function.

Despite this variation, they all have two things in common: they have a soil that is at least periodically saturated or covered with water and they contain plants that can tolerate such conditions (Urban Land Institute, 1985). These wetland conditions are measurable by three criteria: the presence of hydrophytic vegetation, hydric soils, and wetland hydrology. Making these conditions into operational definitions for regulations, administration, and biological mapping is another matter. If FWS's national wetland maps (begun some 14 years ago) are an example, the mapping function has been relatively straightforward.

Table A.3 illustrates this problem for land owners affected by regulations including various federal and state definitions. Biologically or scientifically, they are very similar except that Connecticut's definition includes all poorly drained soils. Only in the legal process can any real difference between these definitions be found (Urban Land Institute, 1985). As part of the 1977 amendments to the Federal Water Pollution Control Act of 1972, the EPA and the COE became responsible for implementing the new wetland provisions in Section 404 of the Act. This legal complexity is represented by the treatise entitled *Law of Wetland Regulation* (Want, 1991). It includes 13 chapters, 152 pages on federal wetland law and procedures, and 196 pages devoted to state wetland law. Much of the book cites judicial opinions, associated procedural permitting, and mapping requirements. The presenting of cases before the Supreme Court further confuses this issue (Want, 1991). However, because the Clean Water Act did not include an explicit definition or procedure for field identification of wetlands, a team of wetland biologists from the FWS, COE, EPA, and SCS in 1989 established common field procedures. The collaborative group combined the best procedures from existing manuals and developed some new procedures to assist in identifying the upland edge of wetlands. These new procedures were designed to include (1) all wetlands regulated by COE and EPA under Section 404 of the Clean Water Act; (2) all wetlands administered under the Food and Security Act of 1985 (Swampbuster); and (3) wetlands mapped by the FWS's NWI.

These criteria include a range of (a) permanently flooded to seldomly flooded, (b) aquatic systems to terrestrial systems, or (c) areas where water dominates to where upland dominates. Somewhere along that gradient, science and society say "That's a wetland" (Seligmann and Hager, 1991). At the federal level, the issue remains: Where does one draw the line? What is or what is not a wetland?

TABLE A.3. Seven Examples of Wetlands Definitions

Emergency Wetlands Resources Act of 1986 (P.L.99-645)
The term wetland means land that has a predominance of hydric soils and that is inundated or saturated by surface or groundwater at a frequency and duration sufficient to support, and that under normal circumstances does support, a prevalence of hydrophytic vegetation typically adapted for life in saturated soil conditions.

Swampbuster Provision, Food Security Act of 1985 (P.L.99-198)
The term wetland, except when such term is part of the term converted wetland, means land that has a predominance of hydric soils and that is inundated or saturated by surface or groundwater at a frequency and duration sufficient to support, and that under normal circumstances does support, a prevalence of hydrophytic vegetation typically adapted for life in saturated soil conditions.

U.S. Fish and Wildlife Service (Cowardin *et al.*, 1979; adopted 1980)
Lands transitional between terrestrial and aquatic systems where the water table is usually at or near the surface or the land is covered by shallow water. For purposes of this classification, wetlands must have one or more of the following three attributes: (1) at least periodically, the land supports predominantly hydrophytes, (2) the substrate is predominantly undrained hydric soil, and (3) the substrate is nonsoil and is saturated with water or covered by shallow water at some time during the growing season of each year.

U.S. Environmental Protection Agency (40 CFR 230.3, Federal Register, 1980) *and the U.S. Army Corps of Engineers* (33 CFR 328.3, Federal Register, 1982)
Those areas that are inundated or saturated by surface or groundwater at a frequency and duration sufficient to support, and that under normal circumstances do support, a prevalence of vegetation typically adapted for life in saturated soil conditions. Wetlands generally include swamps, marshes, bogs, and similar areas.

State of Wisconsin (NR 115.03 WAC)
Those areas where water is at, near, or above the land surface long enough to be capable of supporting aquatic or hydrophytic vegetation and that have soils indicative of wet conditions.

State of Connecticut (22a-38 Connecticut General Statutes) Wetlands means land, including submerged land, which consists of any of the soil types designated as poorly drained, very poorly drained, alluvial or flood plain by the National Cooperative Soils Survey, as may be amended from time to time by the Soil Conservation Service of the U.S. Department of Agriculture.
State of California (California Coastal Act of 1976, Section 30121) Lands within the coastal zone that may be covered periodically or permanently with shallow water and include saltwater marshes, freshwater marshes, open or closed brackish marshes, swamps, mudflats, and fens.

As noted earlier, before this federal interest some states assumed responsibility for the regulation of wetlands. By the mid-1970s additional states also became interested in the protection of wetlands. Even though federal law imposes national consistency, wetland protection has been increasingly augmented by state law (Want, 1991). At present, 27 states have some form of wetland law: explicit wetland regulations, regulationsincluded in coastal zone management, or regulations included in other natural resource management provisions, such as shoreline, beach, and sand protection (Want, 1991).

Regulation of wetlands and their associated definitional requirements has become a complex regulatory arena with numerous judicial decisions interpreting these regulations. This has occurred at all levels of government (Want, 1991). As part of this process of regulation and protection, attorneys, environmentalists, realtors, corporate professionals, scientists, and planners have become involved in the definition, regulation, management, alteration, and restoration of wetlands.

INFORMATION REQUIREMENTS

What Are the Information Requirements?

Obviously the requirements for a wetland information system depend on how wetlands are defined. This definitional task is a scientific, social, legal, and political task. Definitions need to be agreed upon in both the

TABLE A.4. Some Principal Wetland Data Sources

Type of Data Source	Information Displayed	Scale	Suggested Wetland Uses
National Wetlands Inventory Maps (FWS, USGS)	A wide variety of information pertaining to vegetation, water regime, and other variables	1:24,000 1:100,000	• Regulatory mapping • Aid in processing permits • Acquisition • Siting
Topographic maps (7½′ and 15′) (USGS)	Topographic contours, major roads, railroads, utility lines, contours, water bodies, houses, town names, county and town boundaries, and vegetated and non-vegetated wetlands	1:24,000 1:62,500	• Interim wetland map • Watershed boundaries • Source of topographic information
Soil survey (SCS)	Soil types	Range from 1:7,200 to 1:15,840	• Soil suitability for onsite waste disposal • Determination of soil structural bearing capacity
State wetland maps (individual states)	Wetland vegetation boundaries. Varied (depending on state)	1:2,400 to 1:24,000	• Interim wetland regulation maps • Permanent wetland maps (depends on scale)
Flood hazard boundary maps (USGS, HUD, FEMA)	Flood-prone areas	≈ 1:24,000	• Assess flood-hazard potential at wetland sites

Type of Data Source	Information Displayed	Scale	Suggested Wetland Uses
Floodplain information reports (COE)	Standard project flood-plain, 100-year flood evaluation wetland boundaries (some maps)	Range from 1:500 to 1:12,000	• Assess flood hazard potential at wetland sites
Hydrologic investigations, atlas, hydrology, and water resources (USGS)	Maps differ and may contain: wells, test holes, bedrock, and groundwater quality information	≈ 1:24,000	• Determine groundwater flow systems • Determine acquifer recharge areas
Subdivision Maps (local mu-nicipalities)	Dimensions of property, size, and location of house, width of ease-ments. Wetland and floodplain boundaries (some circum-stances)	1:480 1:720 1:1,200	• Determine precise wetland boundaries (occasional) • Evaluate individual developments
Air photos (USGS, USDA, ASCS, states, private)	Existing uses, vegetation, water resources, roads, etc.	Range from 1:7,200 to 1:58,000	• Define wetland boundaries based on vegetation • Use as image maps • Evaluate individual proposed uses

Aadapted from Burke *et al*. (1985).

political and regulatory worlds. A consistent approach to wetlands is needed to scientifically define, classify, and delineate wetlands and to resolve the issue of what subset of wetlands are to be regulated. This subset does not necessarily represent all wetlands. Good wetland information policy needs to be aimed at ensuring that one definition of wetlands emerges with a clear statement of the subset to be regulated. Today's public policy maybe seen as limiting 10 years from now because of continued losses, new understanding, or further reductions in ground-water and surface-water quality.

Modern information systems and their content must be flexible enough to sustain new societal views and analyses, such as the concept of no net loss. Without a definitional context and long-term perspective, it is not possible to explicitly assess the role and usefulness of information and mapping technology. This array of potential sources, scales, and types of data to determine wetland boundaries is complex, diffuse, and disparate (Table A.4). In addition to the regulatory need for data and information for permits and boundary mapping, data and information are needed for trend analysis, compliance monitoring, and management. These have been defined by the Domestic Policy Council-Interagency Wetlands Task Force (Nelson *et al.*, 1990) (Table A.5).

What Is the Status of Wetland Data and Information?

The status of wetland data remains an open question. The Domestic Policy Council-Wetland Inventory Workgroup was asked to address three related information questions:

- What types of inventories are now being done?
- What type of inventories are needed (i.e., national, regional, local) and why?
- How should the federal government coordinate existing (or new) inventory programs?

The answers to these three questions were provided (Nelson *et al.*, 1990) in the form of 11 recommendations, which were rank ordered (see Table A.6, column 1). These recommendations represent three types of information: large-scale boundary determinants for Section 404 permits; statistical samples for trend analysis; and entity imaging and mapping for inventories such as the NWI.

IMPEDIMENTS TO A NATIONAL WETLAND INFORMATION SYSTEM

Wetlands have become a major issue in the environmental debate. Their state and condition have gained bipartisan congressional interest and federal agency attention. The debate over the Swampbuster provisions of the 1985 and 1990 farm bills (FSA and FACTA) are examples. The more recent debate over the provisions of the *Federal Manual for Identifying and Delineating Jurisdictional Wetlands* (Federal Interagency Committee for Wetlands Delineation, 1989) is yet another example. Initially this manual attempted to identify the upland/wetland boundary of **all** wetlands of management interest to COE and EPA (jurisdictional), USDA (Swampbuster), and FWS (NWI).

Through public hearings conducted by the federal agencies and by the President's Domestic Policy Council during the summer and fall of 1990, many expressed their belief that the text of the 1989 federal manual was too encompassing. The procedures incorporated into the 1989 effort did not resolve the definitional problem because public and private interests did not support the protection and regulation of all wetlands of federal interest. Instead of a consensus being arrived at concerning what subset should be regulated, the procedures were challenged. This challenge resulted in the definitions being changed. The issue becomes one of which wetlands are to be regulated. Technically the delineation procedures appeared to be acceptable to most scientists, but farmers and land developers were not comfortable. The real issue became one of the societal value of wetlands. At what point on the wetness gradient does a wetland fall below the public's interest?

This resulted in the proposed 1991 revisions to the *Federal Manual for Identifying and Delineating Jurisdictional Wetlands*. The proposed revised document establishes policy for those jurisdictional wetlands that are proposed to be regulated under Section 404 and Swampbuster. If implemented as proposed, the revisions would move the line delineating the wetland upland boundary toward the wetter end of the moisture gradient, thereby removing drier-end wetlands from federal jurisdiction. This change in boundary would result largely from a revised quantitative standard for the duration and timing of the presence of water.

Even though explicit mandates exist for all wetland management categories, it is now not possible to assemble or assimilate a composite

TABLE A.5. Relevant Legislation and Authorities for Federal Agency Wetlands Mapping and Inventory Functions

Agency	Authority	Agency Function
DOI/FWS	Emergency Wetlands Resources Act of 1986	1. Produce national wetlands inventory maps for the conterminous United States by Sept. 30, 1998 2. Produce ASAP after 1998 maps of the United States in non-conterminous areas 3. Produce by Sept. 30, 1990, and in ten year intervals after that, updated reports on "Status and Trends of Wetlands and Deepwater Habitats in the Conterminous United States" 4. In 1989 produce the report "Wetlands Losses in the United States."
USDA/SCS	Food Security Act of 1985 (FSA) and Food, Agriculture, Conservation, and Trade Act of 1990 (FACTA)	Through the Swampbuster provision, convert wetlands to agricultural production
	Section 302 of the Rural Development Act of 1972	Provide a National Resource Inventory (NRI) including data on the status, condition, and trends of soil, water, and related resources; data for the 1987 NRI were based on more than 300,000 randomly sampled sites; wetland presence is one element of the NRI
USDA/FS	National Forest Management Act of 1976 and Resources Planning Act of 1974	Inventory and manage National Forests
DOC/NOAA	Magnuson Fishery Conservation and Management Act of 1976 with amendments of 1986	1. Identify and describe the habitat requirements of fish stocks 2. Identify existing habitat conditions and sources of pollution and degradation 3. Conduct habitat protection and enhancement programs 4. Recommend measures to protect and manage habitat (fishery habitat includes emergent wetlands, mangroves, and seagrass beds)

Agency	Authority	Agency Function
DOC/NOAA	Coastal Zone Management Act of 1972, reauthorized in 1985 and 1990	1. Plan comprehensively for and manage development of the nation's coastal land and water resources 2. Provide coastal zone enhancement grants to coastal states for protecting, restoring, or enhancing the existing coastal wetlands base or creating new coastal wetlands
DOI/FWS and DOC/NOAA	Clean Water Act, Sec. 404 and Fish and Wildlife Coordination Act of 1958	NOAA and FWS review and comment on S.404 wetland permit applications regarding potential fishery habitat impacts and ways to avert them (fishery habitat inventories are critical to this review process)
EPA	Section 404 of the Clean Water Act	1. Conduct mapping through site-specific enforcement actions and "advance identification, which is designed as an anticipatory regulatory approach conducted on regional or watershed scales 2. Requires states to report on the status of their wetland resources (S.305[b])
	Comprehensive Environmental Response, Compensation, and Liability Act (CER-CLA) and Superfund Amendments and Reauthorization Act (SARA)	Conduct mapping through site-specific mapping nearby or adjacent to National Priority List sites
DOI/FWS and other federal agencies	OMB Circular A-16 on "Coordination of Surveying, Mapping, and Related Spatial Data"	Coordinate national digital mapping of wetlands with the involvement of the federal, state, and local governments as well as the private sector (enable data transfer between producers and users, through data standards)

Adapted from Nelson *et al*. (1990, p. 9-10).

national view. Let us explore in more detail what impedes the ability to develop a national view of wetlands.

As previously stated, our objective was to document the status of spatial data products and associated information technology that support decisions concerning the state and condition of the nation's wetlands. Specifically, the committee discovered a set of the major impediments to the creation of a national wetland information system:

- *Technical impediments*: What are the nature and type of technical and scientific impediments that inhibit the ability to identify, collect, classify, automate, and integrate wetland data on a national basis?
- *Legislative impediments*: What are the legislative mandates that limit or inhibit the development of information systems and the ability for public agencies and private groups to exchange information? What are the legal stipulations that limit authority to implement a national wetland information system of sufficient reliability and specificity for assessment, management, and regulation?
- *Institutional impediments*: What are the existing institutional impediments that inhibit the ability of local, state, and federal agencies and private interests to collect and integrate wetland formation to formulate a national perspective? How do various disciplinary approaches affect the definition and classification of wetland and regulatory issues?
- *Economic impediments*: What are the funding constraints that inhibit the ability to maximize public and private information investments for the purpose of understanding the nature and condition of U.S. wetlands?

This study was conducted under the assumption that these impediments exist in various forms and intensities. Effective use of the public's investment in wetland information is limited by these impediments.

We also assumed that these impediments can be related to other spatial data layers as well. These impediments transcend both political boundaries and institutional structures. The MSC's aim was to improve its understanding of the major issues involved in the creation and population of a wetland data layer as part of the National Geographic Data System (NGDS) (FGDC, 1991). This includes institutions and organizations collecting data and information that would be used to develop such a data base. It also includes institutions and organizations using the data and information for assessing the extent, condition, and regulation of wetlands. The results are intended as an example of the more generic problems associated with the creation and use of other data layers in the NGDS.

To assess the viability of the 11 recommendations of the Domestic Policy Council-Interagency Wetlands Task Force, each recommendation was analyzed with respect to each impediment. The results suggest that various impediments must be overcome for successful implementation of each recommendation (Table A.6).

The ability to create and share a digital version of the NWI (recommendation 7, Table A.6) is constrained by four technical impediments. Integrated and across agency nationwide analyses to determine the status of wetlands, such as acreage per wetland type, are restricted because of duplicative and nonintegrated efforts (legislative and institutional impediments; recommendation 2, Table A.6). Until recently, comprehensive development of a digital wetland layer was limited to the availability of funds directly from users (an economic impediment). No funds were available from the FWS. However, in the view of the DOI's Office of the Inspector General (1992) the most constraining impediment was that the FWS did not have authority to automate the NWI. The accounting process required by the DOI Office of the Inspector General relegated digital conversion to special project status. No such mandate was provided under the Emergency Wetlands Resources Act (EWRA) of 1986 (P.L.99-645). Recent congressional action (P.L.102-440, Section 305) provided that "by September 30, 2004, a digital wetland data base for the United States based on the final wetland maps produced under this section" and "archive and make available for dissemination wetland data and maps digitized under this section as such data and maps become available."

Another major impediment is the ongoing debate and differences of opinion over the reliability and validity of various wetland mapping techniques and procedures. Those responsible for the NWI have concluded that satellite image sources in themselves are not sufficiently accurate to detect all the categories within the NWI system (FGDC, 1992; Wilen and Pywell, 1992). Those responsible for FSA determinations have found problems with aerial photographic techniques due to interpreter differences.

Other professional interagency debate exists when NWI and FSA wetlands are compared. Because the NWI and FSA definitions are operationally the same, statistical and spatial concurrence should be attainable. Recent comparisons in Indiana (SCS, 1992) and the pothole region of North Dakota suggest otherwise (Margaret Maizel, personal communication, 1992). Because the FWS chose not to map wetlands in

TABLE A.6 Impediments to a National Wetlands Spatial Data Infrastructure

Recommendations (Wetlands Inventory Report)	Technical Impediments	Legislative Impediments	Institutional Impediments	Economic Impediments
1. Complete NWI maps (FWS)	• Nonautomated product • No image backdrop (not essential but would enhance product usefulness) • Existing satellite resolution limits usefulness	• No authority to use FWS resources to automate (does not impede 1986 emergency mandate but restricts pace of automation and robustness)	• Other federal, state, and local agencies are duplicating wetland products	• Increased funding required
2. Integrate statistical analyses (FWS, SCS)		• Duplicative and nonintegrated efforts limit full value of wetland quantity and quality data collected	• EPA's EMAP is not included[a]	
3. Implement wetland change program for coastal wetlands (NOAA)	• Limited automated and compatible data sets restricts implementation	• Authority limited to wetlands associated with coastal areas	• Absence of other agency interests (e.g., FWS, SCS, and EPA) results in duplication	
4. Coordinate and integrate national wetland permit tracking system (COE-RAMS)	• Relationship to GRASS is not operational		• Long-term relationship with state permitting is unclear	• Increased funding required

Recommendations (Wetlands Inventory Report)	Technical Impediments	Legislative Impediments	Institutional Impediments	Economic Impediments
5. Develop automation standards for wetland information (FWS)	• Usefulness of SDTS is unclear	• No standard setting authority exists for automated wetland data	• Success of FGDC not assured	
6. Coordinate and integrate wetland mapping programs (FWS, SCS)	• Lack of compatibility between wetland data bases (1:24,000, NWI) in cartographic form with SCS (1:7,920) photographic base[b] • Interface with NWI maps and the SCS orthophoto program are not clear	• No legislative requirement to reconcile and integrate collection platforms, classification systems, and data sets		
7. Establish a National Wetlands Digital Data Base (FWS)	• Lack of automated data sets limits analysis • Inclusion of grade B data • Mixing data sources quickly • Maintaining attribute integrity	• Until recently, no authority to use FWS resources to automate		• Until recently, only available resources for automation were 100% user-pay dollars
8. Expand mapping and inventory systems to include functional value of wetlands (EPA, NOAA)	• Absence of reliable predictive models limits usefulness • Absence of low-cost high-resolution satellite imaging limits applicability		• Relationship with other mapping and statistical interests is duplicative (e.g., SCS, NWI)	

Recommendations (Wetlands Inventory Report)	Technical Impediments	Legislative Impediments	Institutional Impediments	Economic Impediments
9. Establish a national orthophoto program (1:12,000) (SCS/ASCS/USGS)	• Access policy to all image base products (hard and soft) by other federal, state, and local agencies not formulated	• No explicit authority exists to implement such a program	• Limited institutional interest	• Increased funding required
10. Coordinate large-scale digital wetland information (USGS)	• Actual mechanisms for assimilating large-scale data sets into small-scale data sets not yet developed	• No authority exists for agency responsibility	• Limited experience in assimilating data from non-federal sources	• No incentives for data sharing
11. Establish a national digital land cover program (USGS)	• Information technology has not been adapted to implement such a program • Classification system remains confused between land cover (reflective) and land use (activity) • Unit of resolution unresolved	• No explicit authority exists for such a program	• Relationship with proposed national orthophoto program is duplicative • Limited experience in assimilating data from non-federal sources	Increased funding required

[a] For example, others have concluded that . . . "For maximal usefulness (EPA's design) must be adopted by as many federal agencies as possible" (BEST/WSTB, 1992).

[b] Proposed orthophoto program offers a solution (see item 9). Also SCS is investigating satellite imaging systems for FSA wetland determinations.

farmed areas in Indiana, the difference may be explainable. Some differences may be also attributable to the sensor employed. In North Dakota satellite imagery was used to detect FSA wetlands and compared with NWI aerial photographic interpreted sources. This variance in results between NWI and FSA interpretations needs concerted interagency attention. Without its resolution, any real progress towards a national wetland information system will be impeded.

The potential for a national view of wetlands as represented by the 11 recommendations is also constrained by the apparent lack of an overall strategic interagency plan. Such a strategic approach could be as follows:

- Integrate and coordinate NWI (FWS) and FSA (SCS) wetland mapping programs and nest COE jurisdictional wetlands within NWI and FSA (recommendation 6, Table A.6).
- Integrate statistical analyses of the status and trends of the nation's wetlands (recommendation 3, Table A.6).
- Expand FWS's mapping mandate to include automation (recommendation 7, Table A.6) (estimated cost to complete and automate the NWI is about $55 million).
- Develop digital standards for wetland information (recommendation 5, Table A.6).
- Implement the National Orthophoto Program (both digital and hard copy products) (recommendation 9, Table A.6).

Such a strategic approach could be the full implementation of the proposed (SCS/ASCS/USGS) National Orthophoto Program (NOP). An excellent example of how this could be accomplished is Maryland's Digital Orthophoto Quarter Quad Mapping and Wetlands Inventory Program. Maryland's digital quarter quads (1:12,000) incorporate the accuracy and image detail of orthophotography so that land owners and regulators can compare wetland boundaries with recognizable ground features (see Figure A.3). The product meets USGS orthophoto production specifications and the NWI's wetland mapping interpretation and classification standards. See insert for a more detailed description.

This joint effort between Maryland and the FWS is an excellent example of spatial data sharing. It is also an excellent example of a multipurpose and contemporaneous digital information product that is useful for both regulation and planning.

FIGURE A.3 Digital orthophoto (1:7200) inventory for wetlands through a cooperative project by Maryland and the U.S. Fish and Wildlife Service.

MARYLAND'S DIGITAL ORTHOPHOTO QUARTER QUAD MAPPING AND WETLANDS INVENTORY PROGRAM

The Maryland Water Resources Administration (WRA) is producing a statewide map series based on digital orthophoto quarter quadrangles (USGS 3.75' Series). The purpose of the project is to provide a map accurate base image for a new wetland inventory. Orthophotos were selected for the map base because they combine accuracy and image detail so that property owners and permit reviewers are able to see the wetland boundaries in relation to recognizable features on the ground. In order to take advantage of new computer mapping technology, WRA is producing the orthophotos digitally and in color.

There was no precedent for this type of mapping when the project was conceived in 1989. A pilot was conducted with assistance from Salisbury State University, Photo Science, Inc., and Micro Images, Inc. The pilot, performed on the Millington Quad in Kent County, proved sufficiently successful for the state to proceed with the project. With the help of many state and local agencies, federal agencies, universities, and the private sector, the first maps were produced in December 1991. At this writing, 107 quarter quads are completed out of approximately 950 required for statewide coverage.

The orthophotos produced by Maryland are intended to be compatible with the standards of the U.S. Geological Survey for digital orthophoto quarter quadrangles (3.75' Series). The photography used is color infrared flown to the specifications of the National Aerial Photography Program. Geodetic control is acquired and elevation data are collected to produce orthophotos that meet National Map Accuracy Standards at the scale of 1:12,000. The digital image files have a ground resolution of 4 feet and each file occupies approximately 28 megabytes of disk space.

A new statewide wetland inventory is being performed using conventional stereo interpretation techniques with the same color infrared photography that is used to produce the orthophotos. The wetlands are delineated and classified according to the Cowardin *et al.* (1979) classification scheme. Field verification is performed at approximately five sites in each quarter quad, and quality control is provided by the National Wetlands Inventory (NWI) of the U.S. Fish and Wildlife Service. The interpretation is transferred to the orthophoto by NWI and WRA using an on-screen digitizing technique developed in Maryland that results in a more accurate representation than conventional transfer methods. The final wetland map is produced showing the wetland vectors overlaid on the orthophoto image at a scale of 1:7,200.

There are many potential uses of the orthophoto images beyond wetland mapping. Applications in forest inventory, agricultural practices, watershed management, soils mapping, and contour mapping (using the elevation model) are being investigated. Other state agencies, as well as federal and county agencies, are also looking at using the orthophotos for a variety of purposes.

The orthophotos will be integrated into a statewide geographic information system for natural resource management. A prototype GIS has already been implemented by the Water Resources Administration using SPOT Satellite images in 7.5 minute quad format as the base layer. A number of natural resource feature overlays are linked to the base images, and can be accessed in a user friendly atlas format. Regulatory permit data bases can also be displayed geographically in the atlas, and information can be queried and viewed on screen for any individual permit. Eventually, all regulatory boundaries and protected features, such as wetlands, floodplains, critical areas, historic buildings, or endangered species habitat will be able to be viewed and analyzed geographically in one place. This system will save time for permit reviewers who now have to compile mapped information form many sources for a single project, and increase their effectiveness by providing the tools for a more thorough evaluation of impacts, including cumulative effects of other projects. The system will also be useful for studies of regional or statewide scope, such as watershed management plans, wetland mitigation site inventories or regulatory trend analysis. (From Burgess, 1992).

Much of the potential for multiple uses of a national georeferenced and automated wetland data base (see Table A.7) cannot be realized with-out a national perspective. It is the view of the MSC that the creation and implementation of a national wetland information structure is being impeded by more than a lack of funding. For example, the integration and reconciliation of statistical analyses proposed by the FWS and SCS (NRI) need to include the EPA's EMAP, the NOAA, and the USFS. Integration and reconciliation require leadership and may also require legislative action if institutional barriers cannot be surmounted. This analysis also suggests that the Interagency Task Force must adopt a more strategic approach in which various recommendations are coupled rather than simply rank ordered. The FGDC may wish to provide the leadership for this strategic process.

Besides the technical and political impediments, there are institutional issues. Congressional and legislative interest in wetland protection and management has evolved into a complex web of potentially overlapping mandates and authority. Analytical procedures to assess the location, status, and condition of U.S. wetlands and their regulation varies by scale and technique ranging from case-by-case analysis for regulatory assessment functions to trend analyses for determining the status of wetlands, to

TABLE A.7. Uses of a National Georeferenced Wetland Data Base

Determine status/baseline information

- Determine areal extent of vegetated wetlands by type, size, and geographic location
- Determine length in miles of coastal wetlands by type and political subunits
- Determine frequency of occurrence of wetlands by type, size, and location (relative abundance and scarcity)
- Quantify interface between wetland types
- Determine proximity (what is next to what)
- Describe shoreline characteristics (rock versus marsh)
- Establish baseline from which to measure changes

Monitoring change

- Quantify wetland losses and gains by type, size, geographic location
- Quantify wetland modification
- Determine effectiveness of regulatory programs
- Quantify cumulative extent of wetland loss, gain, and modification over time

Provide a tool for wetland management

- Advanced identification or unsuitability determinations
- Flood insurance (FEMA) determinations
- Swampbuster determinations
- Special area designations such as State Heritage Programs
- Risk analysis (oil spill sensitivity)
- Impacts of sea level/climate change
- Mitigation
- Hurricane/storm assessment
- Landscape factors
- Evaluate permitted activity on quality and quantity
- Identify areas subject to development pressure

Determine biotic factors (in conjunction with other data)

- Wildlife habitat potential
- Fisheries habitat potential
- Rare/threatened communities
- Sensitive communities
- Commercial uses such as shellfish/fishing
- Sport uses such as hunting, fishing, bird watching, etc.
- In with other data, quantify gains and losses in functional value (i.e. fisheries production as a function of interface between wetlands and open water

After Nelson *et al.* (1990).

comprehensive entity mapping to determine both locational and distributional characteristics and acreage assessments and status by wetland types. Compounding the overall analytical process, a variety of public and private geographical information technologies are being used to assist and meet the mapping and analysis requirements (Table A.8).

There are also economic impediments. The creation and management of a national wetland information system is not without cost. Reconciling different technologies and approaches should result in some savings. Assimilation of state and local data within the system might, in the long term, provide savings. When we consider what is already being invested to collect wetland information nationwide, the amount is not trivial. For example, in 1991 the COE issued 15,990 wetland permits. Just the private costs to meet the permitting requirements for Section 404 by the COE has been estimated to be more than $100 million annually (Niemann, 1992). This estimate does not include associated costs of litigation that sometimes are also incurred by land owners. A more spatially robust and reliable information system such as that being implemented in Maryland would be helpful in reducing the uncertainty about jurisdictional wetlands and thereby reducing the costs associated with the permit process. Reconciling and integrating NWI and FSA mapping and automation efforts also seem likely candidates for more efficient and effective use of tax dollars.

Wetlands as a potential data layer in the NGDS exemplify spatial and trend information that are dynamic and of major consequence to many in our society. The information and the resultant analysis are of consequence to those in agriculture who farm or develop land and resources, such as real estate developers and oil and mining companies; to those who are concerned with the inherent and functional value of wetlands, such as Ducks Unlimited, The Nature Conservancy, and the Conservation Foundation; and to those who have mandated management and regulatory responsibility for wetlands, such as the FWS, the EPA, the SCS, and the COE.

If we add the annual cost of COE permitting expenditure, the NWI expenditures, the activities of other federal agencies, and state and local mapping efforts, it appears that a more robust solution is economically and technically possible. If this is correct, the question becomes one of leadership. Who is going to seek the authority to create the more robust product? Who is going to establish the various institutional agreements?

INFORMATION DIFFUSION AND EVOLUTION

To address our goal of understanding the current status of spatial data products that support decisions concerning the nation's wetlands, it is important to understand the evolutionary context of wetland information. The specific wetland model arrived at by the MSC is similar to other general information diffusion models. The description of an evolutionary model could serve those who will become responsible for other natural phenomena of national attention. Figure A.4 portrays an information diffusion model that explains and predicts the state and condition of wetland information. This information model is different from a data base because the data eventually serve as a basis for decision making. An example of this evolutionary model is the development of county soil maps prepared by the SCS during the 1930s. Their need became apparent in the dust bowl days when Congress decided that it was in the national interest to combat soil erosion and increase farm productivity. Soil maps became an explicit component of this overall strategy. Soil classification and definition and their attendant functional attributes are now used for a variety of purposes well beyond soil erosion and farm planning.

An example is the use of the hydric soil attribute as an input into the determination of the Swampbuster provisions of the 1985 FSA. States have also mandated the use of soil maps. Wisconsin explicitly requires the use of soil maps for a variety of land planning and management mandates, including state wetland determinations. This evolution from a data base for individual use by farmers for farm crop planning and voluntary soil erosion mitigation planning to regulating use of land based on the prior natural condition (i.e., hydric soils that are evidence of prior wetland condition) is an example of how a data base evolves from a data stage to an information or decision-support stage. This evolution begins with awareness of the problem and ends with regulation. Conceptually this evolution consists of five stages (see Figure A.4).

Awareness Stage

The awareness stage is an ongoing process and includes building a constituency in support of public interests. The case for protection of wetlands has been long and arduous. In this process of debate, considerable amounts

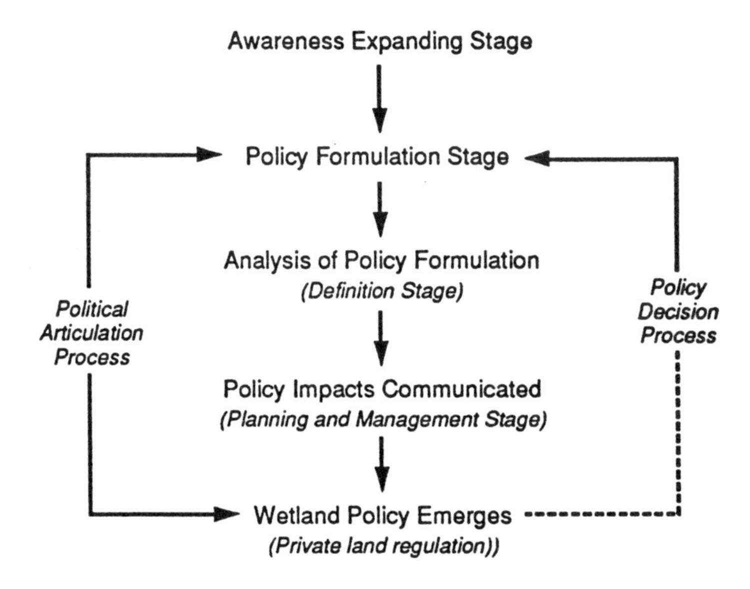

FIGURE A.4. Evolution of wetland information technology diffusion.

of data have been assimilated and converted into descriptive information by ecologists, lawyers, and public interest groups (Kusler, 1983; Office of Technology and Assessment, 1984; Conservation Foundation, 1988). This information tends to be nominal or descriptive and has limited analytical and regulatory value (Table A.9). The basis for this concerted effort was the passage of the 1977 amendments to the Federal Water Pollution Act of 1972 (P.L.92-500). In Section 404 of this act, wetlands associated with navigable waters were protected from discharge and dredge materials (Maxted, 1990).

Policy Formulation Stage

The policy formulation stage establishes legislative intent. Congress expanded its interest and authority in wetlands by the 1977 amendments to the Federal Water Pollution Act of 1972 that established EPA's and COE's authority to include all wetlands as part of the waters of the United States. Waters of the United States are defined in the National Pollution Discharge

TABLE A.8 Analytical Process to Meet Wetland Mandates

Agency/Unit	Mandate	Case by Case	Trend Analysis	Comprehensive Entity Mapping	Spatial Analysis (GIS/LIS)
Department of Agriculture Forest Service (USFS) Soil Conservation Service (SCS)	NFA (1976) RPA (1974) FSA (1985) FACTA (1990) NRI	√	√ NRI	√ 1:24,000 √ 1:7,920 aerial photo base √ 1:15,840 hydric soils √ experimenting with Landsat TM and SPOT	GRASS
Department of Commerce-NOAA Coastal Ocean Program National Ocean Service Coastal Zone Management	MFCMA 1976		√	√	TYDAC ARC/INFO GIS
Department of Defense Corps of Engineers (COE)	CWA S.404	√	√ RAMS		GRASS
Department of the Interior Bureau of Reclamation Bureau of Land Management (BLM) Fish and Wildlife Service (FWS) NWI	 EWRA 1986	√	√ √ report on status and trends Congressionally mandated	√ 1:24,000 NWI √ varies √ 1:24,000 NWI	Provides digital NWI products

Agency/Unit	Mandate	Case by Case	Trend Analysis	Comprehensive Entity Mapping	Spatial Analysis (GIS/LIS)
Environmental Protection Agency (EPA) Advance Identification Program	CWA S.404	√			
Environmental Monitoring and Assessment Program (EMAP)			√		
States					
Wisconsin (as an example, includes 26 additional states, see Want, 1991))	NR 115			√ 1:24,000 image base	ARC/INFO
Private/Non-Profit Ducks Unlimited (DU)		Has discontinued its habitat mapping efforts			

Elimination System (40 CFR Part 122.2) and Section 404 program (40 CFR Parts 230.3 and 232.2) (Maxted, 1990).

As mentioned earlier, states have also taken responsibility for wetland protection associated with their public trust responsibilities. For example, in 1980 Wisconsin as part of its Wisconsin Shoreland Management Program (N.R. 115) extended its land-use zoning authority to include wetlands associated with its streams, rivers, and lakes. The information associated with this stage is initially ordinal in that wetlands are now legislatively different from other lands (e.g., uplands). To assist in the policy formulation phase, status and trends of the nation's wetlands are now being tracked by the FWS and reported to Congress on a 10-year cycle. This initiation of a systematic sample of wetland environments established the ability to conduct statistical manipulations of the data base.

TABLE A.9 Evaluation of Information Technology Diffusion

	Awareness Building Stage	Policy Formulation Stage	Definitional Formulation Stage	Planning Management and Analysis Stage	Private Land Regulation Stage
Type of information	Data	Trends (legislation)	Rules (administrative)	Entities (small scale)	Entities (large scale)
Type of measurement and analysis	Descriptive	Classification	Statistical	Case by case entity mapping prediction	Case by case entity mapping prediction
Level of analysis	Nominal	Ordinal	Ordinal Interval	Ordinal Interval	Ordinal Interval Ratio
Type of information technology used	Maps Photography	Comuter aided drafting functionality Relational functionality	Relational functionality	GIS functionality	Land information system (LIS) functionality

Definitional Formulation Stage

The regulatory definitional stage requires the establishment of administrative rules and the formulation of reliably measurable definitions. The results of this process have a major impact on the data and information content required to implement the states' public policy. Examples of this definitional process includes *Classification of Wetlands & Deepwater Habitats of the United States* (Cowardin *et al.*, 1979); *Corps of Engineers Wetland Delineation Manual* (Wetlands Research Program, 1987); *Wetland Identification and Delineation Manual* (Environmental Protection Agency, 1987), and the *Federal Manual for Identifying and Delineating Jurisdictional Wetlands* (Federal Interagency Committee for Wetland Delineation, 1989). Trend analysis continues to help refine the information available on the status and trends of the nation's wetlands and to continue policy debate.

Planning, Management, and Analysis Stage

The implementation of legislative and Congressional intent and the associated data requirements is the focus of the planning, management, and analysis stage. Mapping of wetland entities becomes a useful means by which to communicate the location and distribution of these regulated and nonregulated wetlands. Examples include entity mapping such as the NWI and continued authority for trend analysis as part of the EWRA of 1986 (P.L.99-645, Title IV, Wetlands Inventory and Trend Analysis, Sec. 401 National Wetlands Inventory Project).

Also in this stage, land-holding agencies are assessing and managing wetlands according to their individual mandates. These include state departments of natural resources, local government interests, and federal agencies such as the USFS, the BLM, Bureau of Reclamation, and National Park Service. Use of GIS technology becomes a major analytical and management tool during this stage.

From the viewpoint of planning, management, and analysis, these trend and entity data collected by the FWS as part of the EWRA are used to establish the base line for determining the actual abundance or scarcity and the rate of conversion of wetlands (Tiner, 1984). Consequently, this analysis of trends forms the basis for much of the public debate over wetlands. For example, the trends data were the predominant source used by the Council of Environmental Quality in its Environmental Trends Report on wetlands and wildlife (CEQ, 1989). Because of the value of the data, the monitoring cycle and the reporting cycle to Congress as mandated by the EWRA needs to be reduced (e.g., 5 years versus 10 years) through reconciliation with the agencies involved in statistical trend and status analysis. It would also be valuable to intensify the national sampling and to produce statistically significant regional estimates on the status and trends of wetlands.

Private Land Regulation Stage

This final stage, private land regulation, is that of imposing stated public policy on privately held lands. Information requirements become more specific and more demanding. Entity mapping becomes integrated with other data such as property. GIS technology and automated land records integration become important factors. Increased analytical capability is expected. This evolution of wetland information diffusion is in a state of flux (Figure A.5). As the definitional process affects the land development rights on privately held land, the public debate accelerates. This interaction, even though troublesome to the information community, in reality constitutes the implementation process, that process being the difference between the policy of no net loss and the political process of what society and private land owners are willing to endure. It is important that the information diffusion process be understood so that consistent and durable policy leading to an enhanced NSDI be formulated.

Although there have been extensive efforts to define the critical properties associated with wetlands, a politically agreeable decision on the subset to be regulated remains at the center of the wetland controversy. This issue of what is a wetland has become a major public debate because it concerns private land. This controversy is further fueled by the administration's stated goal of no net loss. Where the line from wet to dry is drawn has a major impact on how much wetland exists, how much needs protection, and what constitutes no net loss. This issue is further compounded by those who assert that wetlands serve different functions and that only the most important functions need attention. This is even further compounded by the issue of inherent wetland biological and botanical quality. For example, some wetlands have been invaded by exotic wetland plants, such as Purple Loosestrife. Examples of the debate are represented by the following quotes from a selection of newspapers and magazines.

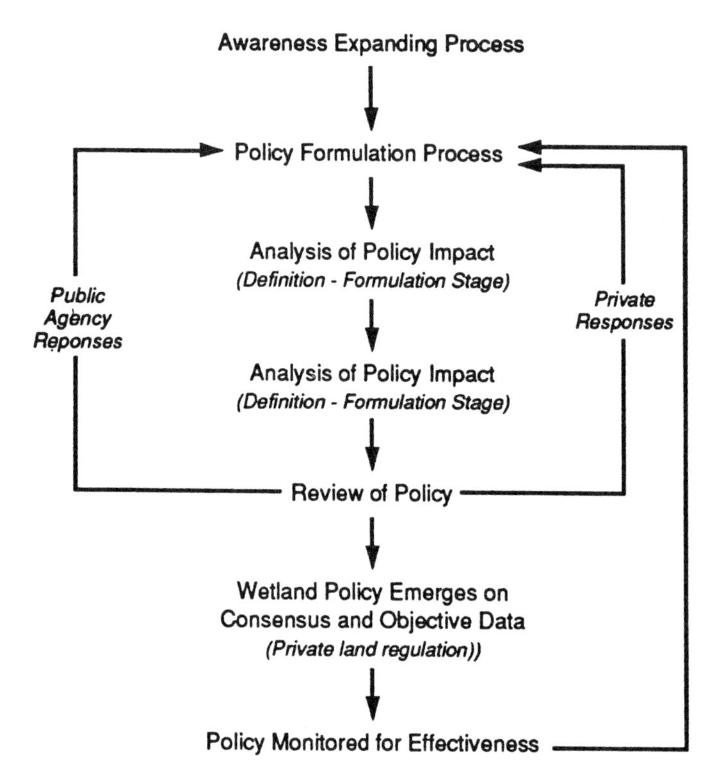

FIGURE A.5 Proposed model for diffusion of wetland information. 'The changes being proposed are political changes not based on good science.' said South Carolina Wildlife Federation executive director . . . (C. Pope, *The State*, August 27, 1991)

A draft of the new document given to the Associated Press by a member of the administration says the previous definition (Interagency Wetlands Manual, Maron, 1990) 'grossly exaggerated' the country's *real* wetlands mostly by not requiring that they be *very* wet. The current definition (March, 1990) says water must come within 18 inches of the surface for at least 7 days of the growing season (this is when chemically soils become hydric in composition and can support hydric vegetation). . . The draft (EPA's) would be stricter, requiring that land be *inundated or saturated* all the way to the surface for at least 14 *consecutive* days in the growing season (*The State*, [Columbia, S.C.] May 15, 1991).

> William Reilly (EPA Chief) thought he had a deal—the Council on Competitiveness (CC)—had agreed that any piece of land that was flooded or saturated for 15 consecutive days a year would constitute a 'wetland' and deserved protection from private development. Within days the council (CC) hatched a new plan, narrowing the definition of 'wetness' by six extra days, satisfying a powerful coalition of farmers and builders and reducing America's wetlands by as much as 30 million acres. (M. Duffy, *Time*, November 4, 1991).

Once the definitional process deviated from agreed-upon measures in the federal manual, the status and condition of wetlands changed, if the national view is limited to those described as jurisdictional. If this limits the national view, then the usefulness of the NWI trends data base as a measure of wetland status becomes questionable. If the perceived national definition remains substantially different from the NWI sample definitions, a new sample design and scheme would be required beginning with the more limited definition. Reconciliation of these issues requires use of a different process to diffuse wetland information. Such a process would need to be responsive to wetland policy and regulatory functions (see Figure A.5). Until we as a nation can agree on the subset of federal wetlands to be regulated, we cannot inventory them or determine whether they are increasing or decreasing.

CONCLUSIONS

We conclude that the need for a common federal view on the location, status, and trends of wetlands and the subset of those of national interest remains as the primary impediment to broad public and private support. This need also impedes the defining of criteria for spatial data about wetlands. For example, the owners of wetlands say "that protection efforts have gone too far" (Zinn and Copeland, 1992). A similar concern was reported in *USA Today* when President Bush was quoted as saying: "We ought to stay with our objective of 'no net loss'. . . but we don't want to overdefine what a wetland is." The debate over wetlands has now moved from a question of whether wetlands should be protected to a question of how much protection should be afforded the remaining wetland resource (Zinn and Copeland, 1992).

A major debate is under way between wetland protection advocates and private land owners. The environmental and wetland advocates have concluded that "a major impediment in maintaining, enhancing, and restoring wetland resources is the lack of a *coordinated, consistent approach* among federal, state and local governments" (Zinn and Copeland, 1992). This lack of a coordinated and consistent inventory mapping and analysis capability is at the crux of public debate. As pointed out by the Executive Director of the Association of Wetland Managers:

> . . . in contrast (to the COE) virtually all state and local governments map wetlands as part of the regulatory process, . . . this lack of a consistent map base results in a federal program subject to varying interpretation by individual regulators. As a result, . . . the regulatory process is difficult and time consuming. . . . Moreover, both delineating a wetland and applying for a permit are costly" (Kusler, 1992, pp. 29-30). (Niemann, 1992, estimated an annual cost of about $100 million.)

Without the reconciliation and interaction of the NWI, the FSA, and jurisdictional wetlands, and Congressional and public support for a composite federal view, the incorporation of spatial data and information about the nation's wetlands into an NSDI remains problematical.

The FGDC, through its Subcommittee on Wetlands, needs to reconcile the definitional and technical issues that impede our nation's ability to efficiently and effectively map, assess, monitor, and automate wetland information. To accomplish this reconciliation, the FGDC needs to exercise its coordination authority (given in OMB revised Circular A-16) and to develop and implement a wetland information-diffusion model that is responsive to both policy and regulatory requirements.

In pursuing this conclusion, the following reconciliation tasks require immediate attention:

- Reconcile what information gathering technologies and combinations are most efficient and effective for completing a national and automated view of the NWI, the FSA, and, eventually, jurisdictional wetlands (i.e., the integration of on-site determinations, aerial photographic interpretation, and satellite imagery detection).
- Reconcile classification and interpretation differences between NWI and FSA wetland delineations (i.e., require that all discernable wetlands are

included in both farmed and nonfarmed regions to ensure the implementation of a composite and scientifically based national wetland data base).

- Reconcile the classification and mapping procedure by which jurisdictional wetland definitions can be nested within the integrated NWI and FSA system (e.g., a separate graphic entity, an attribute to a NWI/FSA wetlands, etc.).
- Reconcile different digital mapping and attribute standards between the NWI, the FSA, and jurisdictional wetlands.
- Reconcile classification and sampling differences between the various statistical status and trends effects being conducted and planned by FWS, SCS, and EPA.
- Reconcile and more precisely define the nature and products of what would constitute a robust national wetland data and information resource (e.g., will it be a traditional NWI cartographic product of 1:24,000? What will be the cartographic representation of wetland entities, polygons, points, or symbols? What will be the cartographic and digital products of the 1:12,000 National Orthophoto Program (e.g., digital or paper products)? Will it be (1) a central automated data base, (2) a distributed digital layer for use by FSA, SCS, EPA, etc., other state and local agencies, private interests such as Ducks Unlimited, or (3) a distributed within-layer data base where every entity provides its part?
- Reconcile how state, local, and federal agencies can provide and gain access to wetland data.

REFERENCES

BEST/WSTB (1992). *Review of EPA's Environmental Monitoring and Assessment Program (EMAP): Interim Report*, Board on Environmental Studies and Toxicology (BEST) and Water Science and Technology Board (WSTB), National Research Council, Washington, D.C., 25 pp.

Burgess, W. S. (1992). Maryland's Digital Orthophoto Quarter Quad Mapping and Wetlands Inventory Program, Maryland Department of Natural Resources, Water Resources Administration.

Conservation Foundation (1988). *Protecting America's Wetlands: An Action Agenda*, The Conservation Foundation, Washington D.C.

Council on Environmental Quality (CEQ) (1989). *Environmental Trends*, Co-sponsored by the Interagency Committee on Environmental Trends, Washington D.C.

Cowardin, L. M. (1982). Wetlands and deepwater habitats: A new classification, *Journal of Soil and Water Conservation.*

Cowardin, L. M., V. Carter, F. C. Golet, and E. T. LaRoe (1979). *Classification of Wetlands & Deepwater Habitats of the United States*, Fish & Wildlife Service, U.S. Department of the Interior, Washington, D.C.

Dahl, T. E. (1990). Wetland Losses in the United States 1780's to 1980's, Fish and Wildlife Service, U.S. Department of the Interior, Washington, D.C.

Dahl, T. E. (1992). Wetland Status and Trends—The Link to Remote Sensing, U.S. Fish and Wildlife Service, U.S. Department of the Interior, St. Petersburg, Florida.

Dahl, T. E., and C. E. Johnson (1991). Status and Trends of Wetlands in the Conterminous United States, Mid-1970's to mid-1980's, Fish and Wildlife Service, U.S. Department of the Interior, Washington, D.C.

DOI Office of Inspector General (1992). *Audit Report: National Wetlands Inventory Mapping Activities, U.S. Fish and Wildlife Service*, Report No. 92-I-790, Washington, D.C., 54 pp.

Ducks Unlimited (1992). Wetland Bacteria Removes Nitrates, Improve Water Quality, Ducks Unlimited 1992 (January/February).

Duffy, M. (1991). Need friends in high places? *Time*, November 4, 1991.

Environmental Protection Agency (1987). *Wetland Identification and Delineation Manual-Volume I Rationale, Wetland Parameters, and Overview of Jurisdictional Approach*, Washington D.C.

FGDC (1991). *A National Geographic Information Resource: The Spatial Foundation of the Information-Based Society*, FGDC First Annual Report to the Director of OMB.

FGDC (1992). Subcommittee on Wetlands working document—Application of satellite data for mapping and monitoring wetlands, Washington, D.C.

Federal Interagency Committee for Wetland Delineation (1989). *Federal Manual for Identifying and Delineating Jurisdictional Wetlands*, U.S. Government Printing Office, Washington, D.C.

Kusler, J. A. (1983). *Our National Wetland Heritage Handbook: A Protection Guidebook*, Environmental Law Institute, Washington, D.C.

Kusler, J., (1992). Wetland delineation: An issue of science or politics, *Environment 34* (2).

Maxted, J. R. (1990). Wetland mapping supported by the U.S. Environmental Protection Agency, *Federal Coastal Wetland Mapping Programs*, Biological Report 90 (18), U.S. Fish & Wildlife Service, Washington D.C.

Nelson, E. H., K. H. Hughes, and R. O. Morgenweck (1990). Memorandum to E.S. Goldstein-Report on the Wetlands Inventory Workshop, dated December 11, 1990), U.S. Department of Commerce, National Oceanic & Atmospheric Administration, National Environmental Satellite, Data and Information Service National Oceanographic Data Center, Washington D.C.

Niemann, B. J., Jr. (1992). Geographical Information Systems (GIS) Technology: Modernizing the Wetland Permitting Process, CRSS Architects, Inc., Houston Texas.

Niering, W. A. (1986). *Wetlands*, The Audubon Society, Alfred A. Knopf, Inc., New York.

Office of Technology Assessment (1984). *Wetlands: Their Use & Regulation*, U.S. Government Printing Office, Washington D.C.

Pope, C. (1991). Wetlands definition blasted, *The State* (Columbia, S.C.), Aug. 27, 1991.

SCS (1992). *Remote Sensing Wetland Recertification Project: Interim Report*, Soil Conservation Service, Washington, D.C., 6 pp.

Seligmann, J., and M. Hager (1991). What on Earth is a Wetland? The White House Seeks a New Definition, *Newsweek*, Aug. 26, 1991.

The State (1991). Proposed "Wetlands" definition draws fire, *The State* (Columbia, S.C.) May 15, 1991.

Tiner, R. W. (1984). *Wetlands of the United States: Current Status and Trends*, U.S. Department of the Interior, Fish and Wildlife Service, National Wetlands Survey, Washington D.C.

Urban Land Institute (1985). *Wetlands: Mitigating and Regulating Development Impacts*, Urban Land Institute.

Want, W. L. (1991). *Law of Wetlands Regulation*, Clark Boardman Co., Ltd., New York.

Wetlands Research Program (1987). *Corps of Engineers Wetlands Delineation Manual*, Technical Report Y87-1, Department of the Army, Waterways Experiment Station, Corps of Engineers, Vicksburg, Mississippi.

Wilen, B. O., and H. R. Pywell (1992). *Remote Sensing the Nation's Wetlands—The National Wetlands Inventory*, U.S. Fish and Wildlife Service, Department of the Interior, Washington, D.C.

Zinn, J., and C. Copeland (1992). Wetland Issues in the 102nd Congress, CRS Issue Brief, Congressional Research Service, The Library of Congress, Washington D.C.

ACRONYMS

ACS	Automated Cartographic System (USGS modernization program)
ANSI	American National Standards Institute
ASCS	Agricultural Stabilization and Conservation Service, USDA
BLM	Bureau of Land Management, DOI
COE	U.S. Army Corps of Engineers
DCW	Digital Chart of the World (DMA product)
DEM	Digital Elevation Model
DFAD	Digital Feature Analysis Data (DMA product)
DIGEST	Digital Geographic Information Exchange Standard
DLG	Digital Line Graph
DMA	Defense Mapping Agency
DOI	Department of the Interior
DOT	Department of Transportation
DPS	Digital Production System (DMA's modernization program)
DTED	Digital Terrain Elevation Data (DMA product)
EPA	Environmental Protection Agency
ESDD	Earth Science Data Directory
EWRA	Emergency Wetlands Resources Act
FAA	Federal Aviation Administration, U.S. DOT
FEMA	Federal Emergency Management Agency
FGCC	Federal Geodetic Coordinating Committee
FGDC	Federal Geographic Data Committee
FHWA	Federal Highway Administration, U.S. DOT
FICCDC	Federal Interagency Coordinating Committee on Digital Cartography

FIPS	Federal Information Processing Standard
FLIP	Flight Information Product (DMA product)
FSA	Food Security Act
FWS	Fish and Wildlife Service, DOI
GBF/DIME	Geographic Base File/Dual Independent Map Encoding File
GIS	Geographic Information Systems
GLIS	Global Land Information System
GNIS	Geographic Names Information System
GPS	Global Positioning System
GRIDS	Geographic Resources Information and Data System (EPA system)
ISDN	Integrated-services networks
ISO	International Standards Organization
IVHS	Intelligent Vehicle-Highway System
LRMP	Land Records Management Program, North Carolina
MARC	MAchine Readable Cataloguing, Library of Congress
MC&G	Mapping, charting, and geodetic data (DMA)
MSC	Mapping Science Committee
NASA	National Aeronautics and Space Administration
NASS	National Agricultural Statistical Service
NAVD 88	North American Vertical Datum of 1988
NAWQA	National Water-Quality Assessment (USGS program)
NCGIA	National Center for Geographic Information and Analysis
NDCDB	National Digital Cartographic Data Base
NGDS	National Geographic Data System
NGRS	National Geodetic Reference System
NGS	National Geodetic Survey, NOAA
NISO	National Information Standards Organization
NIST	National Institute of Standards and Technology
NMD	National Mapping Division, USGS
NOAA	National Oceanic and Atmospheric Administration, Department of Commerce
NREN	National Research and Education Network
NRI	National Resource Inventory, SCS
NSDI	National Spatial Data Infrastructure
NSF	National Science Foundation
OMB	Office of Management and Budget
PLSS	Public Land Survey System (BLM)
PSU	Primary Sampling Units (SCS's NRI)

QA/QC	Quality assurance and quality control
SCS	Soil Conservation Service, USDA
SCSD	Street Centerline Spatial Data Bases
SDTS	Spatial Data Transfer Standard
TIGER	Topologically Integrated Geographic Encoding and Referencing system (Bureau of the Census product)
TVA	Tennessee Valley Authority
URISA	Urban and Regional Information Systems Association
USDA	United States Department of Agriculture
USFS	U.S. Forest Service, USDA
USGS	United States Geological Survey, DOI
USPS	U.S. Postal Service
VPF	Vector Product Format
WAIS	Wide Area Information Servers